Engineering Materials

This series provides topical information on innovative, structural and functional materials and composites with applications in optical, electrical, mechanical, civil, aeronautical, medical, bio- and nano-engineering. The individual volumes are complete, comprehensive monographs covering the structure, properties, manufacturing process and applications of these materials. This multidisciplinary series is devoted to professionals, students and all those interested in the latest developments in the Materials Science field.

More information about this series at http://www.springer.com/series/4288

Joshua Pelleg

Mechanical Properties of Silicon Based Compounds: Silicides

 Springer

Joshua Pelleg
Department of Materials Engineering
Ben-Gurion University of the Negev
Beer Sheva, Israel

ISSN 1612-1317   ISSN 1868-1212   (electronic)
Engineering Materials
ISBN 978-3-030-22600-8   ISBN 978-3-030-22598-8   (eBook)
https://doi.org/10.1007/978-3-030-22598-8

This Springer imprint is published by the registered company Springer Nature Switzerland AG
The registered company address is: Gewerbestrasse 11, 6330 Cham, Switzerland

Dissatisfaction and restlessness are the key
for human progress
Joshua Pelleg

Preface

This book on "Silicides: Mechanical Properties" is unique since all other treatments of this subject appear only in papers and even then only as a part of the general properties of silicides, mainly because of their importance in device fabrication as electronic parts for low-resistivity contacts. The silicides play an important role in very large scale integration (VLSI) as gate material and interconnection in the silicon based integrated circuits not only because of their metal-like low resistivity but also because of their high temperature stability. However, because of their good transport properties and thermal stabilities the silicides are used in thermoelectric technology. Good thermoelectric materials require a large electrical conductivity, high Seebeck coefficient and low thermal conductivity. Transition metal silicides are used as thermoelectric alloys at higher temperatures because of their low toxicity, high natural abundance, good transport properties, and good thermal stabilities. Transition metal silicides are attractive thermoelectrics also because of their mechanical strength, chemical inertness, environmental friendliness with good thermoelectric properties. Further, some refractory silicides, such as $MoSi_2$ were used for heating elements and coatings, as high temperature structural materials due to their high melting point and excellent high temperature oxidation resistance.

High temperature structural silicides represent an important new class of structural materials, with significant potential applications in the range of 1200–1600 °C under oxidizing and aggressive environments.

The first two chapters set the ground of the book by considering the concepts of "what are silicides?" and their structure. Chapter 3 discusses briefly the fabrication methods of silicides, mainly those for the electronic applications. Chapter 4 discusses the use of silicides emphasizing the importance of their mechanical properties. Deformation in silicides either by tension or compression is integral part of Chap. 4, which includes hardness tests as well. It is natural that the closely related topic of dislocation follows the deformation processes providing the understanding why deformation occurs at a stress level much below the theoretical stress. Chapter 5 is devoted to describe dislocations in the silicide considered in this book. Following the pattern set in an earlier book published by Springer that of the Mechanical Properties of Ceramics, Chap. 6 is considering creep (time dependent deformation), followed by cyclic deformation, namely, fatigue, in Chap. 7. Creep is an important deformation process in silicides, as in other materials. Although no theoretical basics have yet been formulated for creep phenomena, leaving those working in the field to rely solely on experimental observations, they should be aware that physical laws govern the complex deformation mechanism in materials exposed to creep conditions. All deformation processes, static, time dependent and cyclic ultimately lead to failure by fracture. Fracture in silicides is an important chapter which intends to provide understanding of fracture and eliminate it as far as possible. Fracture in silicides is discussed in Chap. 8. Mechanical properties of silicides of small dimensions, namely those of nanosilicides are discussed in Chap. 9. Selected alloying is presented in this book, which is an important method to strengthen basic materials, and thus increase the endurance or the time to fracture. An important additive, which can be considered as alloying pivots around B additions to provide various improvements of the various silicides, but mainly its effect on ductility is of great interest. Chapter 10 consists of this important subject. Silicide composites are the subject of Chap. 11. Only some of the composites are discussed in this chapter. Similarly only limited amount of alloyed silicide—considered the most often used— can be included in this book and considered in Chap. 12. Grain size tayloring is an important factor determining the overall mechanical properties of materials. It is Chap. 13 which emphasizes the importance of the dimensions of the grains. The book is closed with the environmental effect in Chap. 14. One cannot conclude a book without summarizing its essential points that the author wants to convey to the audience.

The silides considered in the book are: $CoSi_2$, $NiSi_2$, $FeSi_2$, WSi_2, $TiSi_2$ and $MoSi_2$, which are mostly considered as thin films in microelectronics. In exceptional cases M_5Si_3 will be included due to the absence of pure unalloyed MSi_2. M stands for metal. However, the subjects in this book are considering pure bulk silicides in addition to the silicide film also.

My gratitude to all the publishers and authors for their permission to use and reproduce some of their illustrations and microstructures. Finally, without the tireless devotion, understanding and unlimited patience of my wife Ada, it would be difficult to imagine the completion of this book, despite my decades of teaching the mechanical behaviors of materials. Her helpful attitude was instrumental in inspiring its writing. Here, it is impossible for me not to mention my gratitude to my grandparents for the education they gave me where I spent my childhood and adolescence; they ascended to Heaven in fire, not unlike Elijah the Prophet, though not having been summoned by God.

Beer Sheva, Israel   Joshua Pelleg

Contents

About the Author

Joshua Pelleg received his B.Sc. in Chemical Engineering at the Technion—Institute of Technology, Haifa, Israel; a M.Sc. in Metallurgy at the Illinois Institute of Technology, Chicago, IL; and a Ph.D. in Metallurgy at the University of Wisconsin, Madison, WI. He has been in the Ben-Gurion University of the Negev (BGU) Materials Engineering Department in Beer Sheva, Israel since 1970, and was among the founders of the department, and served as its second chairman. Professor Pelleg was the recipient of the Samuel Ayrton Chair in Metallurgy. He specializes in the mechanical properties of materials and the diffusion and defects in solids. He has chaired several university committees and served four terms as the Chairman of Advanced Studies at Ben-Gurion University of the Negev. Prior to his work at BGU, Pelleg acted as Assistant Professor and then Associate Professor in the Department of Materials and Metallurgy at the University of Kansas, Lawrence, KS. Professor Pelleg was also a Visiting Professor: in the Department of Metallurgy at Iowa State University; at the Institute for Atomic Research, US Atomic Energy Commission, Ames, IA; at McGill University, Montreal, QC; at the Tokyo Institute of Technology, Applied Electronics Department, Yokohama, Japan; and in Curtin University, Department of Physics, Perth, Australia. His non-academic research and industrial experience includes Chief Metallurgist in Urdan Metallurgical Works Ltd., Netanyah, Israel; Research Engineer in International Harvester Manufacturing Research, Chicago, IL; Associate Research Officer for the

National Research Council of Canada, Structures
and Materials, National Aeronautical Establishment,
Ottawa, ON; Physics Senior Research Scientist, Nuclear
Research Center, Beer Sheva, Israel; Materials Science
Division, Argonne National Labs, Argonne, IL; Atomic
Energy of Canada, Chalk River, ON; Visiting Scientist,
CSIR, National Accelerator Centre, Van de Graaf Group
Faure, South Africa; Bell Laboratories, Murray Hill, NJ;
and GTE Laboratories, Waltham, MA. His current
research interests are mechanical properties, diffusion in
solids, thin film deposition and properties (mostly by
sputtering) and the characterization of thin films, among
them various silicides.

Chapter 1
What Are the Silicides?

Abstract In this chapter the silicides are defined. They are the product of the reaction between silicon and metal. Almost all metals can react with silicon forming various silides. Of particular interest are the MSi_2 (M stands for metal) silicides. In this book the $CoSi_2$, $NiSi_2$, $FeSi_2$, WSi_2, $TiSi_2$ and $MoSi_2$ are considered either as bulk or thin films.

Silicides are binary compounds of silicon with other more electropositive elements. Chemical bonds in silicides may exhibit primarily covalent or primarily ionic characteristics, depending on the electronegativity of the participating elements. Similar to borides and carbides, the composition of silicides cannot be easily specified as covalent molecules, since the chemical bonds in silicides range from metal like structures to covalent or ionic. The metal-like structure provides to many of the silicides a conductive character. Of the silicides the transition metal silicides are usually inert to aqueous solutions with exception of hydrofluoric acid.

Silicides are a group of compounds, comprising silicon in combination with one or more metallic elements. These hard, crystalline materials are closely related to inter-metallic compounds and have, therefore, many of their physical and chemical characteristics and some of the mechanical properties of metals. Further they do not appear as natural products and therefore must be synthesized. A metallurgical technique for silicide formation is by directly depositing a refractory metal on a silicon surface to form the required silicide layer by reaction between them. After the metal is deposited on the silicon, this system is exposed to high temperatures that promote the chemical reactions between the metal and the silicon. In such a metallurgical reaction, metal-rich silicides generally form first, and continue to grow until all the metal is consumed. When the metal has been consumed, silicides of lower metal content start appearing, which can continue to grow simply by consuming the metal-rich silicides. At the final point the system attains stability usually when the silicide is a MSi_2, where M stands for a metal. Clearly, enough silicon must be available for silicide formation by the direct metallurgical reaction between metal and silicon. This requirement is generally fulfilled when the metal is deposited on silicon substrate which is sufficient even if the silicon is consumed during the reaction forming the silicide.

© Springer Nature Switzerland AG 2019   1
J. Pelleg, *Mechanical Properties of Silicon Based Compounds: Silicides*,
Engineering Materials, https://doi.org/10.1007/978-3-030-22598-8_1

Almost all metals-definitely a majority-react with silicon. However, not all silicides are of interest for practical engineering and structural applications. The silicides for high temperature applications are limited to the refractory metals of groups IVA, VA and VIA. Included in this category are the silicides of titanium, zirconium, hafnium, vanadium, niobium, tantalum, chromium, molybdenum, and tungsten. However, in the microelectronic industry, the silicides of the VIIIA group metals of the periodic table are of great interest as contacts, gate and interconnection in silicon integrated circuits. Their low resistivity is the main characteristic of these silicides among them, $FeSi_2$, $CoSi_2$, $NiSi_2$, PdSi and PtSi. Silicide classification of elements in the periodic table is presented in Table 1.1 after Murarka.

Various silicides are formed during the process of fabrication when a metal is deposited on silicon. The phases of the silicides form sequentially, but often more than one phase is present which makes the identification of the silicide more difficult requiring skill for the operation of various techniques in addition to X-ray diffraction (XRD), such as Rutherford backscattering (RBS) and Auger electron spectroscopy (AES). In the thin film silicide formation resistivity measurements is an important tool to evaluate the end-product of the silicidation, which exhibits usually the lowest resistivity. Care must be taken—in the processes of thin silicide film formation for the silicon based microelectronic industry—to avoid SiO_2 formation between metal and silicon which inhibits reaction between the silicon and the deposited metal. The misfit between the deposited metal and the silicon introduces stress in the silicide film

Table 1.1 Silicides of elements in the periodic table. Murarka (1995). With kind permission of Elsevier

IA	IIA	IIIA	IVA	VA	VIA	VIIA	VIII			IB	IIB	IIIB	IVB	VB	VIB	VIIB	0
H_4Si																	
$Li_{15}Si_4$ Li_2Si													B_6Si B_4Si B_3Si	CSi	N_4Si_3	OSi O_2Si	F_4Si
$NaSi$	Mg_2Si		$TiSi_3$ Ti_5Si_3 $TiSi$ $TiSi_2$	V_3Si V_5Si_3 VSi_2	Cr_3Si Cr_5Si_3 $CrSi$ $CrSi_2$	Mn_3Si Mn_5Si_3 $MnSi$ $MnSi_2$	Fe_3Si Fe_5Si_3 $FeSi$ $FeSi_2$	Co_3Si CO_2Si $CoSi$ $CoSi_2$	Ni_3Si Ni_2Si Ni_5Si_2 Ni_3Si_2 $NiSi$ $NiSi_2$				Si	PSi	S_2Si	Cl_4Si
KSi KSi_6	Ca_2Si $CaSi$ $CaSi_2$	Sc_5Si_3 $ScSi$ Sc_2Si_3 Sc_3Si_5	Zr_4Si Zr_2Si Zr_3Si_2 Zr_6Si_5 $ZrSi$ $ZrSi_2$	Nb_4Si Nb_5Si_3 $NbSi_2$	Mo_3Si Mo_5Si_3 Mo_3Si_2 $MoSi_2$		Ru_2Si $Ru Si$ $Ru_2 Si_3$	Rh_2Si Rh_5Si_3 Rh_3Si_2 $RhSi$ Rh_2Si_3	Pd_3Si Pd_2Si $PdSi$	Cu_3Si				As_2Si $AsSi$	Se_2Si	Br_4Si
$RbSi$ $RbSi_6$	$SrSi$ $SrSi_2$	Y_5Si_4 Y_5Si_3 YSi YSi_2	$ZrSi_2$												Te_2Si $TeSi$	J_4Si
$CaSi$ $CaSi_3$	$BaSi$ $BaSi_2$	La_5Si_3 $LaSi$ $LaSi_2$	Hf_2Si Hf_5Si_3 Hf_3Si_2 $HfSi$ $HfSi_2$	$Ta_{4.5}Si$ Ta_2Si Ta_5Si_3 $TaSi_2$	W_3Si W_5Si_3 WSi_2	Re_3Si Re_5Si_3 $ReSi$ $ReSi_2$	$OsSi$ $OsSi_2$ $OsSi_3$	Ir_3Si Ir_2Si Ir_3Si_2 $IrSi$ $IrSi_3$	Pt_3Si Pt_2Si $PtSi$							
	••																
•			Ce_3Si Ce_2Si $CeSi$ $CeSi_2$	$PrSi_2$	$NdSi_2$		$SmSi_2$		Gd_3Si_5 $GdSi_2$		Dy_3Si_5 $DySi_2$		Er_3Si_5		$YbSi_x$	Lu_2Si_5
••			Th_3Si_2 $ThSi$ $ThSi_2$	U_3Si_2 USi U_2Si_3 USi_2 USi_3	$NpSi_3$	$PuSi$ $PuSi_2$										

formed, thus the mechanical properties are of interest also in thin film silicides not only in bulk silicide formation. The temperature of the silicide formation is critical because the reaction between metal and silicon is diffusion controlled and as such the diffusion of the metallic atoms into the silicon or vice versa is temperature dependent. It is thus understood why cleanliness of metal-silicon interface is important, since SiO_2 if formed inhibits interdiffusion of the participating entities in the reaction.

As mentioned in the preface of all possible silicides formation the book is considered with the mechanical properties of only the end-phases of a metal-silicon reaction, specifically those of $CoSi_2$, $NiSi_2$, $FeSi_2$, WSi_2, $TiSi_2$ and $MoSi_2$.

Summary

- Silicides are the product of the reaction between metal and silicon.
- The subject of this book are the $CoSi_2$, $NiSi_2$, $FeSi_2$, WSi_2, $TiSi_2$ and $MoSi_2$ silicides.

Reference

S.P. Murarka, Intermetallics **3**, 173 (1995)

Chapter 2
Structure

Abstract In this chapter the structures of the $CoSi_2$, $NiSi_2$, $MoSi_2$, WSi_2, $FeSi_2$ and $TiSi_2$ silicides are presented. Of these silicides $CoSi_2$ and $NiSi_2$ are cubic, $MoSi_2$, WSi_2 and $FeSi_2$ are tetragonal while $TiSi_2$ is orthorhombic. Two phases, C49 (base-centered orthorhombic) and the C54 (face-centered orthorhombic) are of importance, although the C49 variant is metastable. $CoSi_2$ is by far the most important silicide phase because its low electrical resistivity. $NiSi_2$ is also used for device applications. β-$FeSi_2$ is of use in optoelectronics in Si-based devices. Electrical resistivity is the main characteristics of these silicides.

The structure of the above silicides are presented here. They are not the same for all silicides considered. Thus $CoSi_2$ and $NiSi_2$ are cubic, $MoSi_2$, WSi_2, and $FeSi_2$ are tetragonal while the $TiSi_2$ is orthorhombic. Some of these silicides appear as different structures, a good example is $TiSi_2$. The cubic structure of $CoSi_2$ is indicated in Fig. 2.1.

The formation of $CoSi_2$ is by far the most important silicide phase. This is attributed to its low electrical resistivity of ~14 $\mu\Omega$ cm (Tung 2000; Reader et al. 1993) and similar crystal structure to Si. The symmetry of $CoSi_2$ is of type CaF_2 (Muraraka 1983a, b; 1995) and its mostly indicated lattice parameter is a $=5.365$ Å.

Another cubic bisilicide of interest, is the $NiSi_2$, which is the most significant silicide among the Ni–Si alloys for device applications. Its structure is illustrated in Fig. 2.2 showing the bonds between the atoms also. Basically it is the same structure shown for the cubic $CoSi_2$. Although the nickel–silicon system has many phases, only three of those dominate the silicidation process, namely Ni_2Si (resistivity 24 $\mu\Omega$ cm), $NiSi$ (resistivity 14 $\mu\Omega$ cm) and $NiSi_2$ (34 $\mu\Omega$ cm).

After deposition of Ni on silicon, Ni_2Si is formed in the temperature range of 250–300 °C. This nickel rich phase rapidly transforms around 300 °C into the low resistivity $NiSi$ film. However, if the temperature is too high, $NiSi_2$ will appear resulting in a higher electrical resistivity. For electronic device applications the monosilicide $NiSi$ is the most important of the nickel silicides. Its resistivity of 14–20 $\mu\Omega$ cm is comparable to that of $TiSi_2$ and $CoSi_2$, but $NiSi$ can be formed at lower temperatures, namely 350–750 °C. The higher $NiSi_2$ phase is considered in this book in line with the other silicides mentioned in earlier sections. Relevant information on

© Springer Nature Switzerland AG 2019

J. Pelleg, *Mechanical Properties of Silicon Based Compounds: Silicides*,

Engineering Materials, https://doi.org/10.1007/978-3-030-22598-8_2

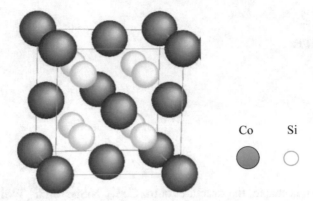

Fig. 2.1 The CoSi$_2$ structure. After Hajjar et al. (2003). With kind permission of Elsevier

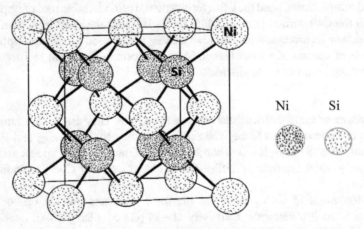

Fig. 2.2 The unit cell of NiSi$_2$ indicating the bonds between the atoms. After Colgan et al. (1983)

the NiSi$_2$ (Jarrige et al. 2009) is: structure C1 (its symmetry is of type CaF$_2$) and its space group is Fm$\bar{3}$m, 4 Ni atoms are in site 4a, and 8 Si atoms are in sites 8c (or the occupancy is Ni 4a and Si 8c); the lattice parameter is a = 5.406 and resistivity 34 ± 2 (Colgan et al. 1983). The majority carriers are electrons in NiSi and holes in Ni$_2$Si and NiSi$_2$.

Figure 2.3 is a common presentation of the tetragonal silicides mentioned above, where M stands for Mo, W or Fe. Thus the unit cell for MoSi$_2$ is shown in Fig. 2.4. Its structure as mentioned is C11$_b$, the space group is I4/mmm, the lattice parameters are a = 3.202, b = 3.202, c = 7.851 (Jarrige et al. 2009). The lattice parameters of the C11$_b$ tetragonal WSi$_2$ (Mattheiss 1992) are a = 3.212 and c = 7.880 (Fig. 2.5). This C11$_b$ ordered-binary alloy is body centered containing three atoms in the unit cell. Three types of atomic layers, one formed by A atoms and two by B atoms, are alternately stacked along the direction of the c axis. The A atoms are the metal and the B atoms are the Si atoms, respectively in the AB$_3$ type tetragonal compounds. Clearly as already indicated above, the space group of the C11$_b$ structure is I4/mmm. Various

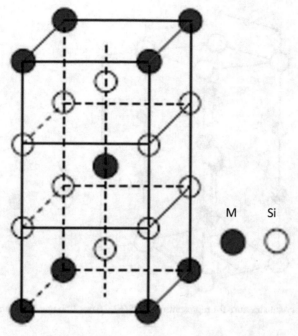

Fig. 2.3 The unit cell of the tetragonal C11$_b$ structure

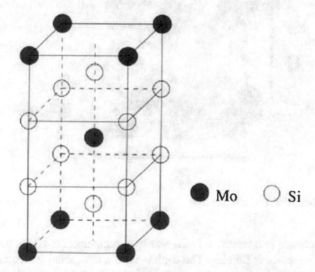

Fig. 2.4 The crystal structure of MoSi$_2$ with the C11$_b$ structure. After Umakoshi et al. (1990)

resistivities for WSi$_2$—from very low to high—were published as listed by Muraraka (1983a, b; 1995) in his Table VIII. In his later review the value of resistivity is given in his Table 2 as 30–70 $\mu\Omega$ cm. The third tetragonal structure is the α-FeSi$_2$. Iron

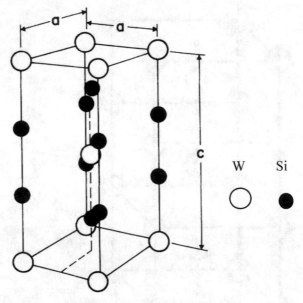

Fig. 2.5 C11$_b$ crystal structure for representing the WSi$_2$. After Itoh and Fujiwara (1992)

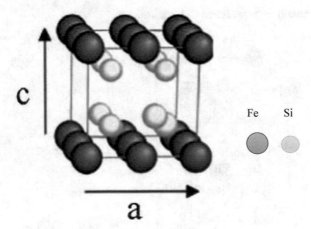

Fig. 2.6 α-FeSi$_2$. After Hajjar et al. (2003)

disilicide occurs in two forms: α-FeSi$_2$ stable at temperatures above 974 °C, while below it the structure is β FeSi$_2$. The α-phase has a tetragonal lattice as mentioned and its unit cell has the dimensions a = 2.69 Å and c = 5.13 Å (Le Corre and Genin 1972) and contains one iron and two silicon atoms. The structure is similar to MoSi$_2$. It is seen in Fig. 2.6 (Hajjar et al. 2003). The resistivity of α-FeSi$_2$ is 230–250 μΩ cm at 4 K and 250–267 at 273 K, thus almost does not change with temperature (Hirano and Kaise 1990). As one unit cell it is illustrated in Fig. 2.7 after Mori et al. (2009).

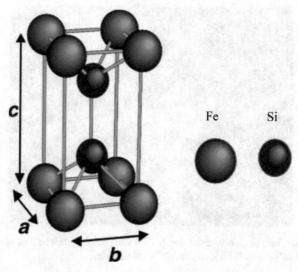

Fig. 2.7 Crystal structure of α-FeSi₂ after Mori et al. (2009). With kind permission of John Wiley and Sons

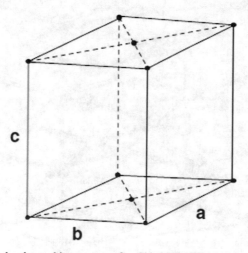

Fig. 2.8 Base-centered orthorombic structure after Chi (2013). With kind permission of Elsevier

Among semiconducting silicides, β-FeSi₂ is the most intensively studied material because of its possible use in Si-based devices in optoelectronics, thermoelectrics and photovoltaics. A base centered orthorhombic structure is seen in Fig. 2.8 which is the crystal structure of β-FeSi₂. A stick and ball model representing the β-FeSi₂ is illustrated in Fig. 2.9.

Its space group is Cmca having 48 atoms per unit cell. The lattice parameters are a = 9.863 Å, b = 7.884 Å, and c = 7.791 Å (Clark et al. 1998). The unit cell has two inequivalent Fe sites, each occupied by 8 atoms as well as two inequivalent Si sites with 16 atoms in each. The α-phase transforms into the β-phase (orthorhombic structure) below ~950 °C according to α FeSi₂ → β FeSi₂ + Si.

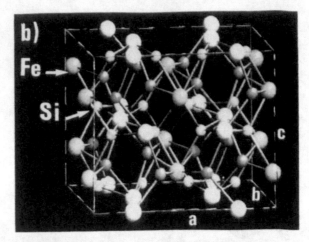

Fig. 2.9 Stick and ball models representing **b** the orthorhombic structure of semiconducting β-FeSi$_2$. Derrien et al. (1992). With kind permission of Elsevier

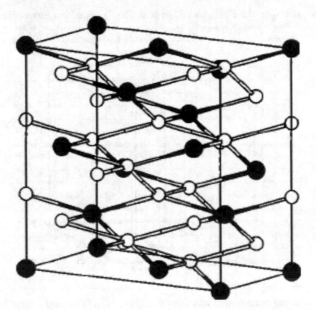

Fig. 2.10 The C54 type crystal structure of TiSi$_2$. The black balls represents the Ti atoms and the white balls represent Si atoms. After Ravindran et al. (1998). With kind permission of AIP Publishing LLC

The TiSi$_2$ formed by solid state reaction of between Ti and Si films has two different structures: the C49 (base-centered orthorhombic) structure formed between 450 and 650 °C, and the C54 (face-centered orthorhombic) structure formed above 650 °C. The C49 TiSi$_2$ is a metastable phase and is usually the first crystalline phase to form in the reaction between Ti and Si, while the C54 TiSi$_2$ phase is a stable phase, having a lower electrical resistivity than the C49 phase. Both crystal structures have

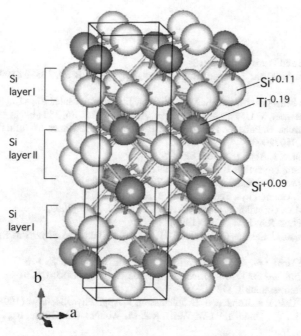

Si layer I

Si layer II

Si layer I

Si$^{+0.11}$

Ti$^{-0.19}$

Si$^{+0.09}$

b

a

Fig. 2.11 Structure of C49 TiSi$_2$. Calculated charges are shown. There are two types of Si layers with differently charged atoms. Ti atoms sandwiched between two types of Si layers, giving a single charged state. Two unit cells are shown. Shudo et al. (2014)

similar atomic arrangements with a hexagonal array of Si atoms around Ti atoms at the center, but the staking arrangement of the unit cells are different: the C49 phase has a two-layer repeat while the C54 phase has a four-layer repeat. C54 TiSi$_2$ has orthorhombic structure with lattice parameters (Ravindran et al. 1998, Niranjan Table 1) of a = 0.8267 nm, b = 0.4800 nm, and c = 0.855 nm and has 24 atoms per unit cell. The space group is Fddd. C49 TiSi$_2$ has lattice parameters of a = 0.355 nm, b = 1.349 nm, and c = 0.355 nm (Niranjan Table 1). Figures 2.10 and 2.11 present the TiSi$_2$ 54 (C54) and the TiSi$_2$ 49 (C49) structures, respectively. As mentioned the structure of TiSi$_2$ consists of close-packed hexagonal layers of composition TiSi$_2$, stacked on top of each other in such a way that Ti atoms of adjacent layers avoid close contact. The periods is of four such layers forming the stacking sequence ABCD in C54, while the C49 has a two-layer repeat. Its space group Cmcm. The thin film resistivities of the C49 and C54 TiSi$_2$ are 52–72 μΩ cm and 2.8–3.8 μΩ cm, respectively (Mammoliti et al. 2002).

Summary

- Th structures of CoSi$_2$ and NiSi$_2$ are cubic, of MoSi$_2$, WSi$_2$ and FeSi$_2$ are tetragonal and that of TiSi$_2$ orthorhombic
- For device applications the resistivity of silicides is of prime importance.

References

D.Z. Chi, Thin Solid Films **537**, 1 (2013)

S.J. Clark, H.M. Al-Allak, S. Brand, R.A. Abram, Phys. Rev. B **58**, 10389 (1998). Beta lattice parameters

E.G. Colgan, M. Mäenpää, M. Finetti, M.A. Nicolet, J. Electron. Mater. **12**, 413 (1983)

J. Derrien, J. Chewrier, V. Le Thanh, J.E. Mahan, App. Surf. Sci. **56**, 382 (1992)

S. Hajjar, G. Garreau, S. Pelletier, P. Bertoncini, P. Wetzel, G. Gewinner, M. Imhoff, C. Pirri, Surf. Sci. **532–535**, 940 (2003)

T. Hirano, M. Kaise, J. Appl. Phys. **68**, 827 (1990)

S. Itoh, T. Fujiwara, Phys. Rev. B **45**, 3685 (1992)

I. Jarrige, N. Capron, P. Jonnard, Phys. Rev. B **79**, 035117 (2009)

C. Le Corre, J.M. Genin, Phys. Status Solidi, (b) **51**, K85 (1972)

F. Mammoliti, M.G. Grimaldi, F. La Via, J. Appl. Phys. **92**, 3147 (2002)

L.F. Mattheiss, Phys. Rev. B **45**, 3252 (1992)

Y. Mori, H. Nakano, G. Sakane, G. Aquilanti, H. Udono, K. Takarabe, Phys. Stat. Solidi B **246**, 541 (2009)

S.P. Muraraka, *Silicides for VLSI Applications* (Academic Press, 1983a), p. 6

S.P. Muraraka, *Silicides for VLSI Applications* (Academic Press, 1983b), p. 31

S.P. Murarka, Intermetallics **3**, 173 (1995)

P. Ravindran, L. Fast, P.A. Korzhavyi, B. Johansson, J. Appl. Phys. **84**, 4891 (1998)

A.H. Reader, A.H. Van Ommen, P.J.W. Weijs, R.A.M. Wolters, O.J. Oostra, Rep. Prog. Phys. **56**, 1397 (1993)

K. Shudo, T. Aoki, S. Ohno, K. Yamazaki, F. Nakayama, M. Tanaka, T. Okuda, A. Harasawa, I. Matsuda, T. Kakizaki, M. Uchiyama, J. Electron Spectrosc. Relat. Phenom. **192**, 35 (2014)

R.T. Tung, J. Cryst. Growth **209**, 795 (2000)

Y. Umakoshi, T. Sakagami, T. Hirano, T. Yamane, Acta Metall. Mater. **38**, 909 (1990)

Chapter 3
Fabrication

Abstract Short description of fabricating silicides is described. Since silicides are used either in bulk or as thin films these processes are considered. Bulk silicides are fabricated by: arck melting and directional solidification, either by Bridgman or Czochralski techniques. Thin films can be produced by sputter deposition, electron beam deposition technique or other deposition methods.

Since silicides are used in thin film form for device production and in various bulk forms, in this chapter short mention of production of silicides will consists of both applications.

3.1 Bulk Silicides

A variety of processing schemes have been used to produce bulk silicides among them arc casting and directional solidification. These methods are briefly discussed below.

3.1.1 Arc Casting (Melting)

Either consumable or nonconsumable vacuum-arc melting can be applied for silicide production. By this technique silicides requiring high melting such as $TiSi_2$, $MoSi_2$ and others can be fabricated. By consumable arc-melting ingots of up to ~7–8 cm in diameter and of ~20 cm in length have been produced. The process essentially is a secondary melting process for production of metal ingots. The alloy (silicides in the present case) to undergo a vacuum arc-melting process can be formed by some consolidation process such as vacuum induction melting. Subsequent consolidation by hot isostatic pressing and/or hot extrusion is often used in the vacuum melted alloy. An electrode in the form of a cylinder is produced under vacuum which is then put into a large cylindrical crucible at the bottom of which a small amount of

© Springer Nature Switzerland AG 2019
J. Pelleg, *Mechanical Properties of Silicon Based Compounds: Silicides*,
Engineering Materials, https://doi.org/10.1007/978-3-030-22598-8_3

the alloy to be remelted is present. The electrode is brought close to prior to starting the melt. A DC current is struck between the two pieces, henceforth a continuous melting process begins. The crucible is surrounded by a water jacket to cool the melt and control the solidification rate. The electrode is lowered as the melt consumes it. Control of the process parameters such as the current, cooling water, the electrode gap etc. is essential to produce the desired quality of the material. However the control of the many variables, such as heat transfer, radiation, convection, etc. is not simple.

In the nonconsumable electrode technique a tungsten electrode is used.

3.1.2 Directional Solidification

Metals or alloys shrink as they change from liquid to solid state. The shrinkage has to be compensated to avoid shrinkage defects in the solidified material. The sprue in castings is a method to avoid shrinkage defects since it is the last portion to solidify being the largest part of the casting. Thus, in a process involving solidification, obtaining sound casting occurs when the solidification proceeds from the furthest end of a casting towards the sprue. The term used is directional solidification. When progressive solidification dominates the directional solidification the shrinkage defect will form in the sprue. The directional solidification process also serves to purify the finished casting product. The basis for the purification is the fact that impurities are more soluble in the liquid than in the solid phase during solidification and the impurities are pushed by the solidification front causing much of the finished product to have a lower impurity concentration than the last solidified part which is enriched with the impurities. This part can be removed from the finished product. In other words in the directional solidification cooling and solidification occur progressively from thin sections to heavy sections with constant metal feed from the heavy section.

An important technique for crystal growth is the float zone processing. The techniques are usable for single crystal growth. In the float zone technique a polycrystalline rod of ultra-pure electronic grade (silicon) is passed through an RF heating coil, which creates a localized molten zone from which the crystal ingot grows. A seed crystal is used at one end in order to start the growth. The whole process is carried out in an evacuated chamber or in an inert gas purge. The molten zone carries the impurities away with it and hence reduces impurity concentration (most impurities are more soluble in the melt than the crystal). Two techniques can be mentioned here, the (a) Bridgman and the (b) Czochralski float zone method.

3.1.2.1   Bridgman

The method involves heating polycrystalline material above its melting point and slowly cooling it from one end of its container, where a seed crystal is located. A single crystal of the same crystallographic orientation as the seed material is grown on the seed and is progressively formed along the length of the container. The process can be carried out in a horizontal or vertical orientation, and usually involves a rotating crucible/ampoule to stir the melt. The technique is a method to produce semiconductor crystal ingots. When no seed crystal is used, the Bridgman float zone technique can be used to obtain polycrystalline ingots.

3.1.2.2 Czochralski

The Czochralski process is a method of crystal growth used to obtain single crystals of semiconductors, metals and other materials. The most important application may be the growth of large cylindrical ingots used in the electronics industry to make semiconductor devices like integrated circuits. The technique involves the slowly pulling upwards while rotating simultaneously a seed crystal which is dipped into a molten pool. By precisely controlling the temperature gradients, rate of pulling and the speed of rotation it is possible to obtain large single crystal cylindrical ingot from the melt.

3.2 Thin Films

Various thin film coating techniques are available. The main two classes are chemical and physical vapor deposition technique. In both techniques one can perform the process either by deposition of a constituent or codeposition of several constituents (namely simultaneous deposition). In the deposition other than codeposition, constituents can be deposited simultaneously and then heat treated to obtain the desired composition. Only some common physical vapor deposition methods are considered in this chapter. Several physical vapor deposition techniques are available to choose from for thin film production. Sputtering and electron beam deposition techniques are briefly described.

3.2.1 Sputtering

In this technique the process involves the ejection of particles from a solid target material by bombardment of the target with energetic gas ions. The most common gas used in sputtering is Ar, but almost any gas can be used to produce energetic gas ions for the bombardment depending on the objective of the process. Argon is mostly used because it's a noble gas. A negative charge is applied to a target source material, A plasma is formed by the ion bombardment. Either DC or RF sputtering cam be used. A figure of merit in DC sputtering is 3–5 kV, while in RF sputtering (alternating current is applied) ~14 MHz is used. The sputter yield (the average number of atoms ejected from the target per incident ion is called the sputter yield) depends on the ion incident angle, the energy of the ion, the masses of the ion and target atoms, and the surface binding energy of atoms in the target. Sputtering can be done from one single target or multiple targets. Sputtering from multiple targets is termed cosputtering. For efficient sputtering magnetron sputtering technique is used. In such process magnetic field is used. Because ions are charged particles, magnetic field can be used to control their velocity and behavior. The ejected atoms arrive to a substrate where the desired film is deposited.

3.2.2 Electron Beam Deposition

It is a physical vapor deposition technique in which an anode (target) is bombarded with an electron beam from a charged tungsten filament under high vacuum. The target atoms become a gaseous phase the atoms of which deposit or precipitate into solid form. Coating everything in the vacuum chamber occurs during the process. The electron beam deposition occurs at a high deposition rate at lower deposition temperatures than the chemical vapor deposition method. The generated electron beam is accelerated to a high kinetic energy directed toward the evaporation material. Upon striking the evaporation material, the electrons loose rapidly their energy. Under proper condition the electron's kinetic energy is converted into thermal energy. The thermal energy that is produced heats up the evaporation material causing it to melt or sublimate. At sufficiently high temperature (and vacuum) the melt or solid transforms to vapor. The resulting vapor then coats exposed surfaces. The thermal energy is obtained by converting the electron kinetic energy (~85%) when the accelerating voltage is 20–25 kV. Usually the electron voltage can be in the range 3–40 kV.

A range of other processing techniques are available but they are out of the scope of this book. One could mention that secondary processing such as forging and extrusion are often applied to finished products by any of the above techniques in order to provide certain mechanical properties to the consolidated materials. But also this secondary processing can be and are used for obtaining the net-shape component or product. Further to obtain good mechanical properties forging or extrusion are used to obtain a desirable micro-texture of the consolidated material by any of the consolidation process. For optimal properties any of the mentioned processes must be carefully controlled.

Summary

- Long ingots are produced by directional solidification
- The directional solidification methods can be used for purification of the product
- Thin films by sputtering are produced by ejection of particles from a solid target or
- By physical vapor deposition involving the bombardment of a target with electron beam from a tungsten filament.

Chapter 4
Testing-Deformation

Abstract This chapter considers the major static deformations observed in silicides. The tests performed are tension, compression and indentation in both single crystals and polycrystalline silicides. Further, the mechanical properties of thin films are an integral part of this chapter and their stress developed during formation on cooling (differential thermal shrinkage) and as a result of mismatch between substrate and silicide film are discussed. Known expressions are included in this chapter which describe the observed stress developed and the strain rate sensitivity. The ideal case to eliminate cracking, other defect formation and lift-off in the silicide film is zero stress. Slip lines of some single crystal silicides for various orientations and temperatures characterize the deformation. Silicides are brittle at room temperature and become ductile at elevated temperatures. The single crystal and polycrystalline silicides and silicide thin films presented in this chapter are $CoSi_2$, $NiSi_2$, $MoSi_2$, WSi_2, $FeSi_2$ and $TiSi_2$.

4.1 Introduction

The mechanical properties are understood in terms of the relation between atoms and the bonding electrons coordinating them. In the silicides considered in this book transition metals of the groups IV, VI and VIII are bonded to a Si atom. The chemical bonds in silicides range from conductive metal-like to covalent or ionic bonding. Due to the metal-like character of silicides, a fundamental characteristics is its electrical resistivity. Other basic properties responsible for the wide use of silicides are their high-temperature stability and excellent corrosion resistance. Further, the transition metal silicides are usually inert to aqueous solutions with the exception of hydrofluoric acid (HF). Some of the silicides considered in this chapter posses outstanding mechanical and other engineering properties, such as hardness and strength which makes them suitable structural materials for application even at elevated temperature due to their high melting point. Potential applications include aerospace manufacturing of parts such as rocket nozzles etc. As thin films, various silicides—some of them included in this book—are used in microelectronics for VLSI applications. In this section mechanical testing of selected silicides is discussed. It is assumed that

© Springer Nature Switzerland AG 2019
J. Pelleg, *Mechanical Properties of Silicon Based Compounds: Silicides*,
Engineering Materials, https://doi.org/10.1007/978-3-030-22598-8_4

the reader is familiar with the basic equations, however the reader can refresh them by consulting the books Mechanical Properties of Materials (Pelleg 2013) and/or the Mechanical Properties of Ceramics (Pelleg 2014).

4.2 Tension

4.2.1 CoSi₂

Surprisingly almost all research activity on deformation in bulk and thin films reports compression and almost no tensile testing is recorded. The reason—at least in thin films probably is a consequence of the fact that during the silicidation reaction two sources of stress arise in the silicide: a compressive intrinsic stress and a thermal stress. The origin of the compressive stress is the volume changes occurring during silicidation and that of the tensile thermal stress due to the difference in the thermal expansion coefficients between the silicide and the silicon substrate. After the formation of the $CoSi_2$ at elevated temperature a thermal stress builds up in the silicide film. The thermal stress is tensile because the thermal expansion coefficient of $CoSi_2$ is higher than that of the silicon substrate. The residual built-in stress at room temperature is 1.1 GPa.

When a silicide is formed on the silicon substrate, usually stress develops in the silicide film. As mentioned the stress consists of two main sources intrinsic and thermal. The total stress can thus be expressed as

$$\sigma = \sigma_\iota + \sigma_T \tag{4.1}$$

The thermal stress can be expressed as

$$\sigma_T = (\alpha_{Msi} - \alpha_{Si})[E/(1 - \nu)]\Delta T \tag{4.2}$$

where α_{Msi} and α_{Si} are the thermal expansion coefficients of the silicide and the Si, respectively, and E and ν are the Young's modulus and the Poisson ratio of the silicide. M stands for the respective metal comprising the silicide, in the present case it is Co. The thermal expansion of a silicide is in the range of $\sim 10 \times 10^{-6}$ K^{-1} and that of Si 2.6×10^{-6} K^{-1} and their difference is sufficient to induce a tensile stress in the silicide as it cools down from the silicide formation temperature. A third stress component might be induced into the silicides depending on the deposition condition. In Fig. 4.1 a tensile stress in $CoSi_2$ is seen as a function of formation temperature.

From Fig. 4.1 the biaxial elastic constant (E/1 − ν) was determined as 140 GPa and for the temperature independent stress a value of 0.8 GPa was derived. This stress was attributed to the temperature independent stress contribution to the intrinsic stress resulting from the volume changes occurring during the silicidation process.

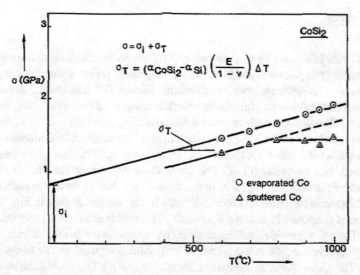

Fig. 4.1 Tensile film stress as a function of silicide formation temperature. The linear behaviour observed for CoSi$_2$ films formed from evaporated Co enables one to distinguish thermal stress (σ_T), due to differences in the expansion, from intrinsic stress σ_i due to volume effects upon silicidation. van Ommen et al. (1988). With kind permission of AIP Publishing LLC

Stress should be reduced in the film which is an important parameter that must be considered. Excessive stress may cause the silicide film to break up and lead even to the loss of adhesion to the substrate. Further the ~1 GPa stress in CoSi$_2$ film may cause to defects—such as pinholes—formation leading to large current leakage and subsequent device failure. Therefore the stress has to be reduced as far as possible. A method to reduce the tensile stress in CoSi$_2$ films formed is by implantation of C ions into the Si substrate before the deposition of Co film on the Si substrate. It was found that the stress in the CoSi$_2$ films decreases linearly with the increase of the C implantation dose (Fig. 4.2).

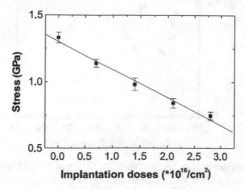

Fig. 4.2 The relationship between the internal stress of CoSi$_2$ films and the C implantation doses in the Si substrates. Liu et al. (2002). With kind permission of Elsevier

4.2.2 NiSi₂

Nickel silicide is a common silicide material for advanced MOS transistors due to its
low formation temperature, low Si consumption, and low line width sheet resistance
dependence as compared to cobalt or titanium silicide. The stresses that develop during the silicidation by the metal-silicon reaction are generally compressive. However,
some data indicate that tensile stresses develop during some silicide formation. Tensile stress is induced in the NiSi₂ film when instead the conventional annealing a two
step process is used which combines the conventional rapid thermal annealing (RTA)
with pulsed laser annealing (PLA). The stress value depends on the silicide thickness
as illustrated for NiSi₂ in Fig. 4.3. In the figure two flow patterns are indicated for
the silicidation in the two step process as seen in the schematic flow in Fig. 4.4. This
technique is proposed to achieve a smooth silicon-rich interfacial layer on (1 0 0)
silicon. The PLA provides sufficient effective temperature during silicidation and
leads to increased tensile stress of the silicide film compared to the two-step RTA
process. The technique is an alternative for the two-step RTA silicidation process for
ultra-scaled devices. It is expected that an improved interface morphology, reduced
interfacial resistance and better ultra-scaled device performance will result. The flow
pattern indicates that Ti capping has been performed after oxide removal. The process is intended to reduce the Schottky barrier height for continued use of nickel
silicide in transistors. It has been indicated that the Schottky barrier height in NiSi₂
single crystal is reduced by 0.3 eV. The sheet resistance is an important parameter in
device performance. The relation between the tensile stress and the sheet resistance
for the three flow patterns indicated in Fig. 4.4 is shown in Fig. 4.5.

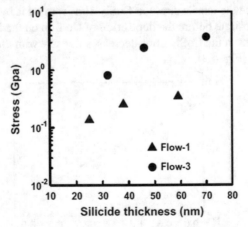

Fig. 4.3 The stress and silicide thickness correlations. Different 1st RTA conditions (300, 350 and
400 °C, all for 15 s) were used to obtain different silicide thickness for the flow-1 and different
PLA energy density (0.6, 1.5 and 2.3 J/cm²) were used to obtain different silicide thickness for the
flow-3. Chen et al. (2010). With kind permission of Elsevier

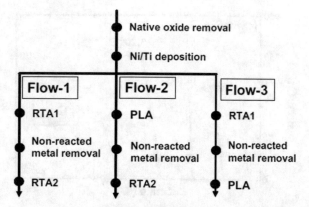

Fig. 4.4 Three different annealing process flow charts for the silicidation evaluation. Chen et al. (2010). With kind permission of Elsevier

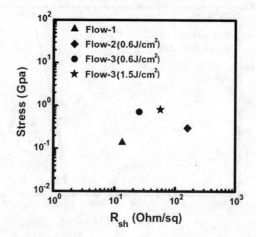

Fig. 4.5 The stress and sheet resistance correlations of different annealing sequences. The PLA energy density of the flow-3 samples were 0.6 and 1.5 J/cm^2. Chen et al. (2010). With kind permission of Elsevier

4.2.3 *MoSi$_2$*

MoSi$_2$ is a promising high temperature material due to its oxidation resistance up to 1950 K and good high temperature mechanical properties. However no plastic deformation occurs below 1200 °C as can be seen from Fig. 4.6. At 1200 and 1300 °C plastic deformation can be seen. The curve means that the brittle–ductile transition temperature occurs at high temperature and at low temperature the single crystal is brittle. This is obvious also from Fig. 4.7 where the elongation in tension and fracture strain in compression as a function of temperature are indicated.

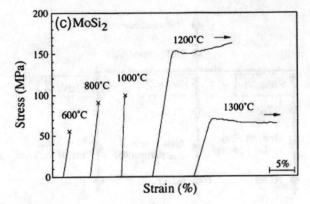

Fig. 4.6 Typical tensile stress-strain curves in MoSi$_2$ single crystals. Nakano et al. (2002). With kind permission of Elsevier

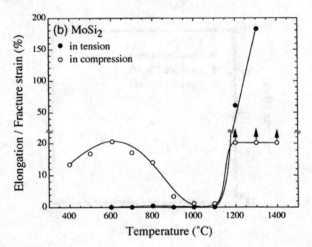

Fig. 4.7 Temperature dependence of elongation in tension and fracture strain in compression for MoSi$_2$ single crystals. Nakano et al. (2002). With kind permission of Elsevier

The temperature dependence of the critical resolved shear stress (CRSS) is seen in Fig. 4.8. In both figures, in that of 4.7 and 4.8 data obtained by compression are included. It might be of interest to show in Fig. 4.9 the loading axis and geometry of the specimens used to collect the experimental data.

4.2.4   WSi$_2$ Film

As known the measured stress in silicide films is composed of two components the intrinsic stress which arises during film formation and the thermal stress, σ_T which

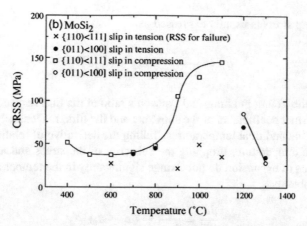

Fig. 4.8 Temperature dependence of CRSS for the {110}⟨111] slip below 1200 °C and the {011}⟨100] slip at above 1200 °C in MoSi₂ single crystals with increasing temperature. Fracture stress in MoSi₂ is described as a cross after being changed into resolved shear stress (RSS) for the {110}⟨1̄11] slip. Nakano et al. (2002). With kind permission of Elsevier

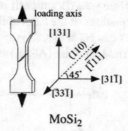

MoSi₂

Fig. 4.9 Showing crystallographic orientations for MoSi₂ and the loading axis. Nakano et al. (2002). With kind permission of Elsevier

is a consequence of the difference in thermal expansions between film and substrate. The film stress if assumed to be isotropic in the plane of the substrate is given by

$$\sigma = \frac{E}{6(1-\nu)} \frac{D^2}{t\Delta R} \tag{4.3}$$

where ΔR is the net radius of curvature, D is the thickness of the substrate, t is the film thickness and E and ν are Young's modulus and Poisson's ratio for the substrate respectively. The values 1.8×10^{12}, 5.0×10^{12} and 8.5×10^{11} dyn cm^{-2} were used for $E/(1-\nu)$ of (100) Si (Brantley 1973) and for (1̄102) sapphire (Wachtman et al. 1960) substrates respectively. The values used for the thermal expansion coefficients of the substrates were 3.2 ppm °C^{-1} for silicon (Burkhardt and Marvel 1969) and 7.7 ppm °C^{-1} for (1̄102) sapphire (Wachtman et al. 1962) in the temperature range 20–400 °C.

The thermal stress is usually expressed as

$$\sigma_T = \frac{E_f}{1 - v_f} \int_{T_1}^{T_2} \left(\alpha_S - \alpha_f\right) dT \tag{4.4}$$

E_f and v are the Young modulus and Poisson's ratio of the film respectively, α_S and α_f are the thermal coefficients of the substrate and the film, respectively and T_1 and T_2 are the initial and final temperatures. Taking the derivative of relation (4.4) with respect to the temperature, dropping the subscript of the stress and assuming that the coefficients of expansion do not change significantly in the temperature range of interest one obtains

$$\frac{d\sigma}{dT} = \frac{E_f}{1 - v_f} (\alpha_S - \alpha_f) \tag{4.5}$$

Figure 4.10 shows the stress-temperature curves for WSi_2 films on silicon and sapphire substrates.

In the absence of any information about either α_f or $E_f/(1 - v_f)$, both values can be obtained by simply determining $d\alpha/dT$ on each of two substrates with known values of α_S and solving two equations of the form of Eq. (4.5) for α_f and $E_f/(1 - v_f)$ simultaneously. Alternatively it can be achieved graphically from Fig. 4.11 for the specific case of WSi_2 on silicon and sapphire (Al_2O_3) substrates. The values obtained

Fig. 4.10 Stress-temperature curves for WSi_2 on silicon and sapphire substrates: ●, ▲, heating; ○, △, cooling. Retajzxyk and Sinha (1980). With kind permission of Elsevier

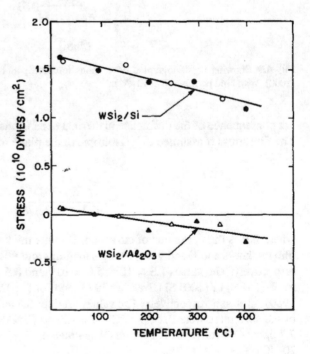

Fig. 4.11 Plots of $\{E_f/(1 - \nu_f)\}^{-1}$ against α_f for WSi_2 on silicon and sapphire substrates. Retajzxyk and Sinha (1980). With kind permission of Elsevier

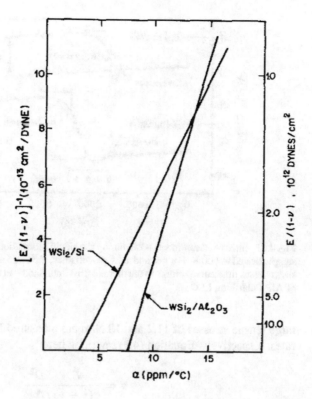

for α and $E/(1 - \nu)$ in the temperature range 20–400 °C are $\alpha = 13.7$ (ppm °C) and 1.2 (10^{12} dyn cm^{-2}), respectively.

4.2.5 FeSi₂ Film

As mentioned earlier iron disilicide is observed in two forms, α-FeSi$_2$ and β-FeSi$_2$. The former is stable below 670 °C and is characterized by a semiconducting behavior, while the latter at >670 °C shows metallic behaviour. β-FeSi$_2$ is orthorhombic. Unfortunately direct reports on the tensile stress in FeSi$_2$ is scarce and mostly the information is indirect. Stress measurements were taken during silicide formation by optical deflection technique as illustrated in Fig. 4.12. The figure is a consequence of continuous measurement of the stress during the growth of Fe films (0.1–1.5 nm thickness) on Si (111) in ultra high vacuum. The deposition up to 0.3 nm Fe induces a compressive stress of -1 N/m. This stress is attributed to the formation of a reactive Fe–Si layer exhibiting a silicide-like structure. Subsequent deposition at 300 K leads to a small tensile stress of 0.7 N/m while deposition at 600 K induces a high tensile film stress of 18 N/m. At a substrate temperature of 600 K a reaction sets in between Fe and Si forming β-FeSi$_2$. As seen from the figure during the silicide formation

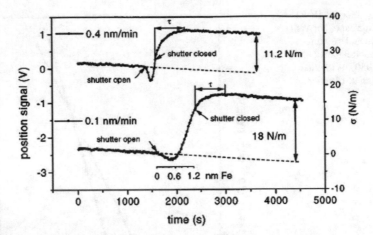

Fig. 4.12 Stress measurements taken during the silicide formation. Fe was deposited on a Si (111) sample heated to 600 K. For fast and slow growth rates, the same amount of time τ passes, until the stress versus time curve returns to the thermal drift line. Sander et al. (1995). With kind permission of AIP Publishing LLC

large tensile stresses of 11.2 and 18 N/m are measured for the high and low growth rates, respectively. Equation (4.3) rewritten here

$$\sigma = \frac{E}{6(1 - v)} \frac{D^2}{t \Delta R} \tag{4.3}$$

was used to calculate the stress from measured sample curvature. The elastic properties of the sample are $E/(1 - v) = 2.23 \times 10^{11}$ N/m^2 for Si (111), the sample thickness is D = 0.15 mm.

High film stresses should be avoided since at high tensile stress of 1300–1600 MPa cracking and delamination is observed.

4.2.6   TiSi$_2$

(a) Film on Polysilicon

As deposited titanium film usually has a relatively low compressive stress up to above 400 °C. On annealing above this temperature the film stress changes to tensile which increases with temperature up to about 900 °C. The meaning of tensile stress is that the film is forced to expand although the tendency would be to contract.

Titanium was deposited on polycrystalline silicon films which were cleaned by a 100:1 (H$_2$O:HF) dip for 2 min followed by a deionized water rinse. The polycrystalline silicon films were obtained by depositing polycrystalline silicon films on oxidized wafers by low pressure chemical vapor deposition process to a chosen film thickness from 2000 to 10,000 Å. The stress measured at room temperature versus annealing temperature is shown in Fig. 4.13 for a specimen where the polysilicon and the film thicknesses were 4300 Å and 1000 Å, respectively. The annealing was carried out in hydrogen (H). In the graph, T and C on the ordinate mean tensile and compressive stress, respectively. As-deposited titanium film has a very low compressive stress which does not change significantly until above 400 °C. Above this temperature on annealing the film has a tensile stress which increases with temperature up to ~900 °C. The stress appears to level off after sintering at temperatures; >1000 °C. Tensile stress in the film would mean that the film is forced to expand (i.e., is being pulled apart) and would like to contract. As more and more silicide grows, there is increased tensile stress.

(b) C49 TiSi$_2$ Film

In Fig. 4.14 stress of a C49 TiSi$_2$ layer on silicon and on sapphire (Al$_2$O$_3$) as a function of temperature is illustrated.

Fig. 4.13 Room–temperature stress in the annealed metal or silicide film (30 min H) as a function of the annealing temperature. Murarka and Fraser (1980). With kind permission of AIP Publishing LLC

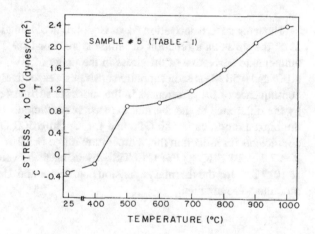

Fig. 4.14 Observed stress in C49 TiSi$_2$ on silicon and sapphire substrates during cooling from 370 to 70 °C (cooling rate: 0.30 °C/min). Jongste et al. (1993). With kind permission of AIP Publishing LLC

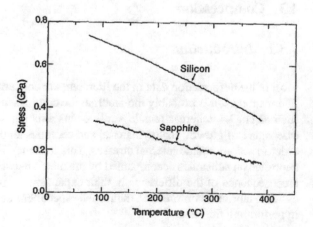

Directly after formation, C49 $TiSi_2$ films exhibit tensile stress. This stress relaxes considerably above 375 °C. From in situ measurements of the film stress as a function of temperature the biaxial modulus $E/(1 - v_f)$ and the thermal expansion coefficient (α) of the film can be deduced. Using Eq. (4.3) the stress can be evaluated form R, the radius of curvature of the sample, and R the intrinsic radius of curvature of the substrate thus providing the value of ΔR. The curvature was measured in situ during cooling in the range from 370 to 70 °C. The cooling rate was 0.30 °C/min. Figure 4.14 shows that the change of the stress is almost proportional to the temperature. The slight deviation from linearity can have various causes such as the variation with temperature of the expansion coefficients and Young's moduli of film and substrates. These were taken as constants in the evaluation of the change in stress with temperature from Eq. (4.4) reproduced here as

$$\sigma_T = \frac{E_f}{1 - v_f} \int_{T_1}^{T_2} (\alpha_S - \alpha_f) dT \qquad (4.4)$$

The temperature derivation $\frac{d\sigma}{dT}$ of Eq. (4.5) is obtained from a linear least squares fit of the stress as a function of temperature. For C49 $TiSi_2$ formed at 470 °C the temperature derivative of the stress in the range of 70–370 °C is 1.40 MPa/K on Si (100) and 0.58 MPa/K on sapphire substrates, respectively. The difference in $\frac{d\sigma}{dT}$ is a consequence of the differences in the substrate stiffness of silicon and sapphire and by the difference in the thermal expansion coefficients between them. (For silicon the biaxial modulus is 180 GPa and for sapphire 500 GPa; the thermal expansion coefficients for silicon in this temperature range is 3.2×10^{-6} K^{-1} and for sapphire it is 7.7×10^{-6} K^{-1}.) For C49 $TiSi_2$ by using these values a value of $10.9(\pm 1.3) \times 10^{-6}$ K^{-1} for the thermal expansion coefficient and 195 ± 12 GPa for the biaxial modulus were obtained.

4.3   Compression

4.3.1   Introduction

Most of the deformation data in the literature are concerned with compression tests. The main reason is probably the fact that the compressing force tend to close flaws such as cracks, whereas tensile stress opens existing small cracks inducing their extension and growth. Brittle material and ceramics tend to produce small cracks in order to accommodate internal stresses. This tendency is magnified when films are deposited on substrates accompanied by mismatch between the film and substrate as a consequence of the difference in their expansion coefficients. Further, it is much easier to fabricate compression than tensile specimens and thus there is a cost saving in performing the tests.

4.3.2 *CoSi₂*

Single and polycrystalline CoSi$_2$ were deformed as a function of orientation in the hope to find some elongation in the otherwise brittle silicide. The expectation of observing reasonable deformability is based on the crystal being of cubic symmetry (C1 structure) and thus with the possibility provided by the lattice of sufficient number of independent slip systems even at ambient temperature. Earlier investigations by the same authors (Ito et al. 1992, 1994, 1999) showed that specimens oriented at orientations different than [001] can deform plastically in compression up to 2–4% strain by slip on {001}(100). The lack of deformability in polycrystals is likely due to the insufficient number of independent slip systems for an arbitrary plastic strain. Only three are independent of the six {001}(100) slip systems, contrary to what is expected from the cubic symmetry of the lattice. CoSi$_2$ polycrystals can however be deformed above 500 °C when secondary slip system other than the {001}(100) primary system becomes operative. At high temperatures {111}(110) and {110}(110) can be operative which can augment the primary slip system.

Figure 4.15 shows some compressive stress-strain curves for single crystals and polycrystals at various temperatures for the orientations indicated. One can see from the curves that the yield stress is temperature dependent for all orientations. Further, it is seen that for orientations [011] and [$\bar{1}$23] where {001}(100) slip systems are

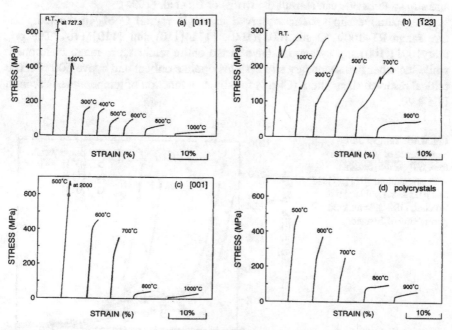

Fig. 4.15 Stress-strain curves of CoSi$_2$ single crystals with orientations **a** [011], **b** [$\bar{1}$23] and **c** [001], and **d** polycrystals at a strain-rate of 1×10^{-4} s^{-1} at various temperatures. Ito et al. (1994). With kind permission of Elsevier

operative plasticity is present even at room temperature. The plasticity in polycrystals is almost the same as that of [001] single crystals, namely the onset of plasticity is above 500 °C. A yield drop is seen in crystals of orientations [011] and [$\bar{1}$23] in the temperature range RT–300 °C which is followed by work hardening. The absolute value of the yield drop decreases with increasing temperature. Similarly, the rate of work hardening also decreases with increasing temperature becoming dominant at and after 800 °C deformation.

The temperature dependence of the yield stress is presented in Fig. 4.16 for the orientations indicated. There is a discontinuous step in each of the yield stress-temperature curves at a temperature between 600 and 800 °C.

The compressive fracture strain of single crystals with the orientations indicated as a function of temperature is shown in Fig. 4.17. The fracture strain of all specimens is seen to rapidly increase with increasing temperature. Table 4.1 lists the Schmidt factors for various slip systems. The Schmid factor for slip on {110}(110) in [$\bar{1}$11]-oriented crystals is essentially zero while that in [011]-oriented crystals is 0.354 as seen in Table 4.1. Slip lines observed in CoSi$_2$ oriented as indicated are seen in Fig. 4.18. Further slip lines observed in CoSi$_2$ are seen in Fig. 4.18. Slip trace analyses on two orthogonal faces were performed on [011], [$\bar{1}$11], [$\bar{1}$23] and [$\bar{1}$35] oriented single crystals. These are deformed to fracture and the results are presented in Fig. 4.19. In the middle column of Fig. 4.19, traces of {001}, {110} and {111} planes on each corresponding face are indicated. For a full analysis of the slip lines and slip systems one can consult the works of Ito et al. (1994).

The operative slip systems observed in single crystal CoSi$_2$ in the temperature range RT–1000 °C are {001}(100), {111}(110) and {110}(110). The primary {001}(100) slip system is active in the entire temperature range mentioned, while the other two secondary slip systems operate only at and above 500 °C. The critical resolved shear stress (CRSS) for slip as a function of temperature is seen in Fig. 4.20.

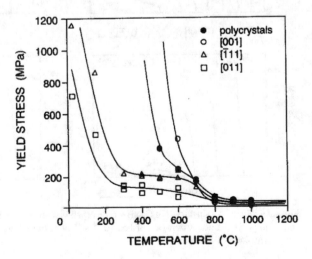

Fig. 4.16 Temperature dependence of yield stress for CoSi$_2$ single crystals with orientations [011], [$\bar{1}$11] and [011], and polycrystals. Ito et al. (1994). With kind permission of Elsevier

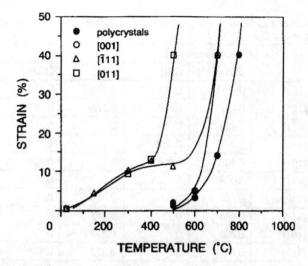

Fig. 4.17 Temperature dependence of compressive fracture strain for CoSi$_2$ single crystals with orientations [011], [$\bar{1}$11] and [001], and polycrystals. Ito et al. (1994). With kind permission of Elsevier

4.3.3 MoSi$_2$

(a) **Single Crystal**

MoSi$_2$ is one of the important silicides (also WSi$_2$, and NbSi$_2$) because it is an extremely high temperature material for use for aircraft, gas turbines and airframes. Due to its excellent oxidation resistance—because the formation of a protective SiO$_2$ film—MoSi$_2$ has been used as a heating element and coating material. In addition the interest in MoSi$_2$ application is associated with its high melting point and high strength. MoSi$_2$ crystallizes below about 1900 °C in a C11$_b$ type ordered structure. Completely disordered MoSi$_2$ has a body centred tetragonal structure and in deformed MoSi$_2$ {110}$\langle 3\bar{3}1 \rangle$ and {013}$\langle 3\bar{3}1 \rangle$ slips coexist.

Compression tests were performed in argon atmosphere at a strain rate of 1.4 × 10^{-4} s^{-1} in the temperature range of 900–1500 °C. The tetragonal structure of MoSi$_2$ illustrated in Fig. 2.4 is reproduced here. The orientation dependence of the observed slip plane is shown in Fig. 4.21. Stress-strain curves (yield stress) for samples deformed by {110}$\langle 3\bar{3}1 \rangle$ and {013}$\langle 3\bar{3}1 \rangle$ slips for the temperatures indicated earlier are shown in Fig. 4.22. The yield stress variation with temperature is seen in Fig. 4.23. It can be seen that the yield stress exhibits a peak (not very sharp) in all orientations tested, which is followed by a gradual decrease of the yield stress with the increase of temperature which is orientation dependent. Except to orientation [001] the yield stress is almost constant at and above 1300 °C for the other orientations.

CRSS for the mentioned orientations are shown in Fig. 4.24. The critical resolved shear stress (CRSS) for {013}-slip is much higher than that for {110}-slip along the

Table 4.1 Schmid factors for various slip systems in $CoSi_2$ single crystals. Ito et al. (1994). With kind permission of Elsevier

Slip system		Compression axis				
		[001]	[Oil]	[Ī11]	[Ī23]	[Ī35]
(111)	[Ī10]	0	0.408	0.272	0.350	0.327
	[Ī01]	0.408	0.408	0.272	0.467	0.490
	[0Ī1]	0.408	0	0	0.117	0.163
(Ī11)	[110]	0	0.408	0	0.175	0.210
	[101]	0.408	0.408	0	0.350	0.420
	[0Ī1]	0.408	0	0	0.175	0.210
(1Ī1)	[110]	0	0	0	0	0.023
	[01 1]	0.408	0	0.272	0	0.093
	[Ī01]	0.408	0	0.272	0	0.070
(11Ī)	[Ī10]	0	0	0.272	0.175	0.140
	[011]	0.408	0	0.272	0.292	0.280
	[101]	0.408	0	0	0.117	0.140
(001)	[100]	0	0	0.333	0.214	0.143
	[010]	0	0.5	0.333	0.429	0.429
(010)	[100]	0	0	0.333	0.143	0.086
	[001]	0	0.5	0.333	0.429	0.429
(100)	[010]	0	0	0.333	0.143	0.086
	[001]	0	0	0.333	0.214	0.143
(110)	[Ī10]	0	0.354	0	0.107	0.114
(1Ī0)	[110]	0	0.354	0	0.107	0.114
(101)	[Ī01]	0.5	0.354	0	0.286	0.343
(Ī01)	[101]	0.5	0.354	0	0.286	0.343
(011)	[0Ī1]	0.5	0	0	0.179	0.229
(0Ī1)	[011]	0.5	0	0	0.179	0.229

entire deformation temperature. The activation of slip $\{110\}\langle 3\bar{3}1\rangle$ and $\{013\}\langle 3\bar{3}1\rangle$ in $MoSi_2$ provides enough slip systems to satisfy the von Mises criterion for continuous deformation. However, the ductility of $MoSi_2$ is not enough below 1200 °C where the anomalous strengthening is expected and the surface markings show coarse and irregular slip bands. Stress-strain variation with temperature is shown in Fig. 4.25 compressed along [201] at temperatures between 300 and 1200 °C at a strain rate of 10^{-5} s^{-1}. Serrations in the stress-strain curve occur as a consequence of the formation of intensely deformed coarse slip bands and not enough plastic strain is expected.

(b) Polycrystals

Molybdenum disilicide polycrystals were deformed in constant strain rate tests in compression at the temperatures indicated in Fig. 4.26. The stress-strain curve show

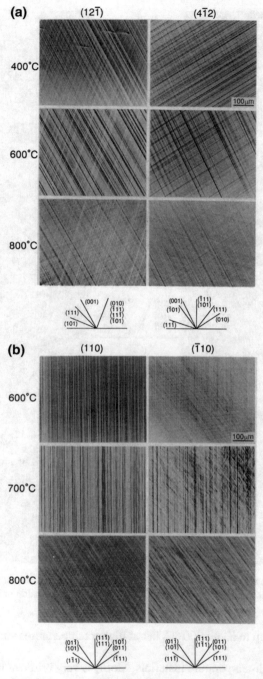

Fig. 4.18 Slip lines observed on CoSi$_2$ single crystals with orientations **a** [$\bar{1}$23] and **b** [011] deformed at various temperatures. The compression axis is in the horizontal direction and traces of {001}, {110} and {111} are indicated in the bottom of the figure. Ito et al. (1994). With kind permission of Elsevier

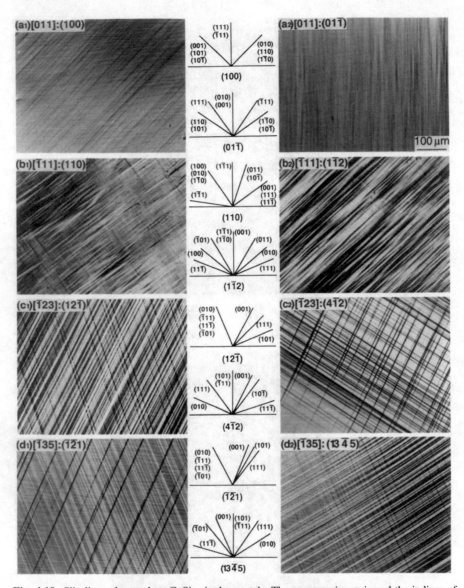

Fig. 4.19 Slip lines observed on CoSi$_2$ single crystals. The compression axis and the indices of the observed surfaces are indicated. Ito et al. (1992). With kind permission of Elsevier

work hardening up to about 900 °C, but at higher temperatures strain softening sets in.

Like other high-temperature materials, MoSi$_2$ is very brittle at room temperature but at high temperatures it becomes ductile. The low strain rate deformation was used since at higher strain rates even at high temperatures the material disintegrates.

The strain rate sensitivity r of the flow stress σ is determined by

Fig. 4.20 Temperature dependence of CRSS for slip on {001}(100), {111}(110) and {110}(110) in CoSi$_2$ single crystals. Ito et al. (1994). With kind permission of Elsevier

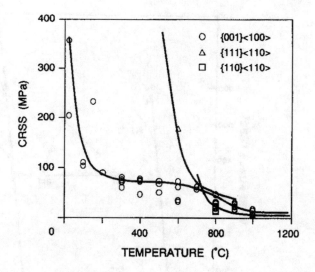

Fig. 4.21 Orientation dependence of the observed slip plane in MoSi$_2$ single crystals deformed at 900 °C. Open circles and open triangles represent {110}- and {013}-slips, respectively. Umakoshi et al. (1990). With kind permission of Elsevier

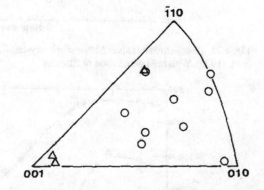

$$r = \frac{d\sigma}{dln\acute{\varepsilon}_{plast}} = \frac{d\sigma}{dln(-\sigma)} \tag{4.5}$$

$\acute{\varepsilon}_{plast}$ is the plastic strain rate which can be determined from the slope of the stress relaxation curves plotted as the logarithm of the relaxation rate versus the stress σ. The strain rate sensitivity is expressed by m given as

$$m = \frac{dln\acute{\varepsilon}_{plast}}{dln\sigma} = \frac{dln(-\acute{\sigma})}{dln\sigma} \tag{4.6}$$

It is expected that grain size affects the stress-strain relation. Fine and coarse grain polycrystals were tested at strain rates different and higher than that indicated in Fig. 4.26. The strain rates are $10^{-3}, 5 \times 10^{-4}$ and 10^{-4}. Figures 4.27 and 4.28 show the load–displacement curves for two grain sizes, namely that of 5 μm and 27 μm and at the mentioned strain rates, respectively. The two MoSi$_2$ materials also differ

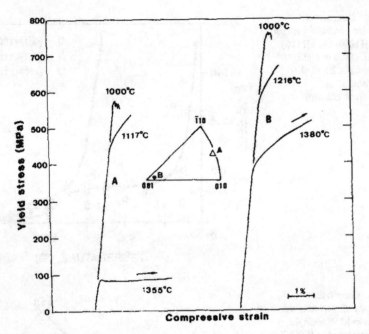

Fig. 4.22 Stress-strain curves of MoSi$_2$ single crystals deformed at various temperatures. Umakoshi et al. (1990). With kind permission of Elsevier

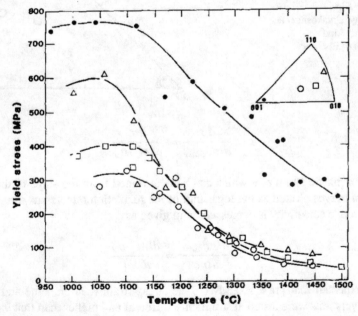

Fig. 4.23 Yield stress of MoSi$_2$ single crystals as a function of temperature. Umakoshi et al. (1990). With kind permission of Elsevier

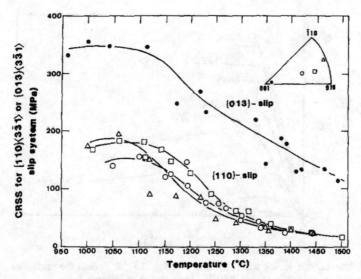

Fig. 4.24 Temperature dependence of the CRSS for $\{110\}\langle3\bar{3}1\rangle$ and $\{013\}\langle3\bar{3}1\rangle$ slip systems. Umakoshi et al. (1990). With kind permission of Elsevier

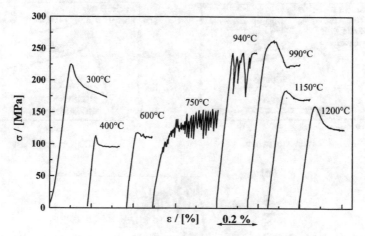

Fig. 4.25 Stress-strain curves of $MoSi_2$ with a [201] compression axis deformed at different temperatures at a strain rate of 10^{-5} s^{-1}. Guder et al. (1999). With kind permission of Elsevier

in their O content. Details of the chemical compositions, impurity content and other pertinent data such as σ, ν and the true strain are listed in Tables 4.2, 4.3 and 4.4, respectively. The variation of the 0.2% (plastic strain) yield stress with temperature is illustrated in Fig. 4.29 for the $MoSi_2$ types. The strain rates were as indicated earlier. As can be seen from the yield strength that the grain size has an effect on the high temperature strength of $MoSi_2$ since the 5 μm material (RHP $MoSi_2$) is higher than that of the one with of 27 μm grain size (Starck $MoSi_2$) when tested up to 1200 °C

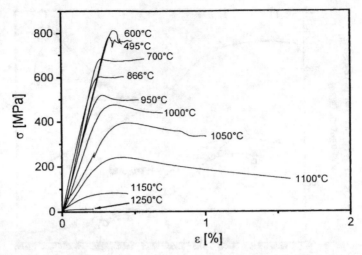

Fig. 4.26 Stress-strain curves at a strain rate of 2.5×10^{-7} s^{-1} until the respective first stress relaxation tests. Junker et al. (2002). With kind permission of Elsevier

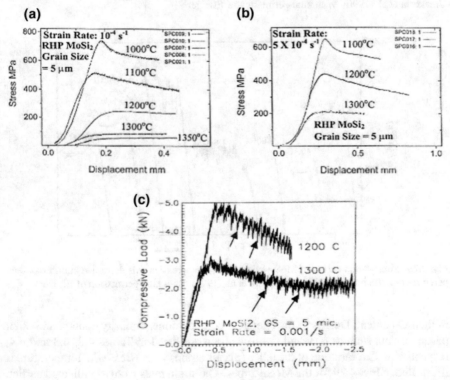

Fig. 4.27 Load displacement or engineering stress displacement curves of RHP MoSi$_2$, corresponding to strain rates of **a** 10^{-4} s^{-1}, **b** 5×10^{-4} s^{-1}, and **c** 10^{-3} s^{-1}. Serrations are shown with an arrow in Fig. 4.27c. Mitra et al. (2003). With kind permission of Springer

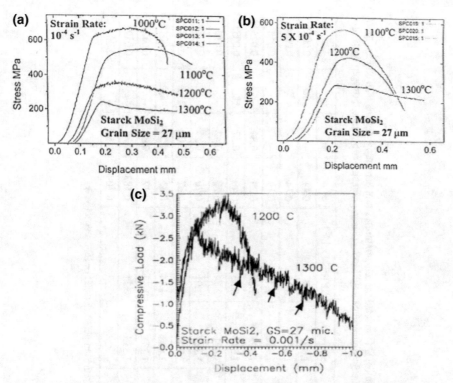

Fig. 4.28 Load displacement or engineering stress displacement curves of Strack MoSi₂, corresponding to strain rates of **a** 10^{-4} s^{-1}, **b** 5 × 10^{-4} s^{-1}, and **c** 10^{-3} s^{-1}. Serrations are shown with an arrow in Fig. 4.28c. Mitra et al. (2003). With kind permission of Springer

Table 4.2 Chemical composition of impurities in Mo, Si, and Starck MoSi₂ powders. Mitra et al. (2003). With kind permission of Springer	Powder (source)	Impurity content (wt pct)
	Mo powder (NFTDC)	W ≈ 0.62, FE ≈ 0.018, C ≈ 0.052, O ≈ 1.3
	Si powder (Johnson Matthey, Inc.)	Fe ≈ 0.21, Mg ≈ 0.01, C ≈ 0.021, O ≈ 0.23
	MoSi₂ (H.C. Starck)	Fe ≈ 0.11, Mn ≈ 0.036, C ≈ 0.08, O ≈ 1.0, N ≈ 0.032

with a strain rate of 10^{-3} s^{-1} and up to 1100 °C when tested at 5 × 10^{-4} s^{-1}. At higher temperatures, however the fine grained MoSi₂ shows a lower yield strength. Further the strain rate sensitivity of the fine-grained MoSi₂ is higher than that of the coarse-grained one, and sharply increases with increase in temperature in the strain rate range of 10^{-3} to 10^{-4} s^{-1}.

The strain hardening also is temperature dependent as seen from Fig. 4.30 where it is plotted versus the true plastic strain and it varies with strain rate and grain size. In Fig. 4.31 the yield stress versus strain rate is shown for the types of MoSi₂, namely

Table 4.3 Values of σ_y, σ_{max}, σ_{max}/σ_y, mean SHR normalized by shear modulus, G (corrected for temperature), strain hardening exponents, n and n_1, strain softening parameter γ_σ, and true plastic strain at maximum stress $\varepsilon_{p,m}$ from the flow curves[a] of RHP $MoSi_2$ (Mean grain size = 5 μm, and oxygen content = 0.06 wt pct). Mitra et al. (2003). With kind permission of Springer

Strain rate (s^{-1})	T (°C) and machine	σ_y (MPa)	σ_{max} (MPa)	σ_{max}/σ_y	SHR/G ($\times 10^2$)	n	n_1	γ_s	$\varepsilon_{p,m}$
10^{-3}	1100 (D)	757	758	1.001	8.0	0.44	–	6.1	0.0025
10^{-3}	1200 (D)	497	594	1.194	0.71	0.14	–	–	0.0095
10^{-3}	1300 (D)	263	313	1.190	0.70	0.19	–	–	0.018
10^{-3}	1350 (I)	180	217	1.205	1.26	0.29	0.84	–	0.0253
5×10^{-4}	1100 (I)	626	652	1.042	5.55	0.40	0.57	10.1	0.0055
5×10^{-4}	1200 (I)	407	440	1.081	2.74	0.28	0.42	–	0.010
5×10^{-4}	1300 (I)	181	214	1.182	1.01	0.20	0.35	–	0.0219
10^{-4}	1000 (I)	740	747	1.009	1.39	0.07	0.08	23.3	0.0047
10^{-4}	1100 (I)	475	517	1.088	4.84	0.31	0.50	8.7	0.0044
10^{-4}	1200 (I)	216	244	1.131	1.38	0.22	0.36	–	0.0132
10^{-4}	1300 (I)	72	94	1.315	0.31	0.183	0.32	–	0.0492
10^{-4}	1350 (I)	42	–	–	0.28	0.209	0.56	–	>0.06

[a]The equipment used, INSTRON or DARTEC, is shown in parentheses next to the temperature as (I) or (D), respectively

Table 4.4 Values of σ_y, σ_{max}, σ_{max}/σ_y, mean SHR normalized by shear modulus, G (corrected for temperature), strain hardening exponents, n and n_1, strain softening parameter γ_s, and true P%plastic strain at maximum stress $\varepsilon_{p,m}$ from the flow curves[a] of Starck MoSi$_2$ (mean grain size = 27 μm, and oxygen content = 0.89 wt pct). Mitra et al. (2003). With kind permission of Springer

Strain rate (s^{-1})	T (°C) and machine	σ_y (MPa)	σ_{max} (MPa)	σ_{max}/σ_y	SHR/G × 10^2	n	n_1	γ_s	$\varepsilon_{p,m}$
10^{-3}	1200 (I)	405	452	1.116	2.32	0.38 0.13	0.48 0.17	–	0.0161
10^{-3}	1300 (D)	329	333	1.011	5.40	0.38	–	–	0.0042
10^{-3}	1350 (I)	322	328	1.018	1.80	0.18	0.33	12	0.0038
5 × 10^{-4}	1100 (I)	492	557	1.132	3.51	0.27	0.42	–	0.0126
5 × 10^{-4}	1200 (I)	407	431	1.059	2.0	0.22	0.40	–	0.0096
5 × 10^{-4}	1300 (I)	285	289	1.014	0.84	0.10	0.12	5.6	0.0042
10^{-4}	1000 (I)	628	672	1.07	1.38	0.13 0.04	0.27 0.15	–	0.0236
10^{-4}	1100 (I)	498	555	1.114	2.0	0.25 0.05	0.61 0.18	–	0.0251
10^{-4}	1200 (I)	330	358	1.086	1.35	0.14	0.44	–	0.0119
10^{-4}	1300 (I)	242	245	1.014	0.73	0.08	0.12	10.2	0.0047
10^{-4}	1350 (I)	146	147	1.007	1.92	0.35	0.79	6.8	0.0027

[a]The equipment used, INSTRON or DARTEC, is shown in parentheses next to temperature as (I) or (D), respectively

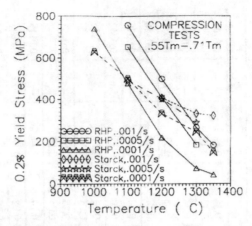

Fig. 4.29 Plots showing variation of 0.2 pct compressive yield stress of RHP MoSi$_2$ (solid lines) and Strack MoSi$_2$ (dashed lines) at strain rates of 10^{-4} s^{-1}, 5×10^{-4} s^{-1}, and 10^{-3} s^{-1} with temperature between 1000 °C (0.55 T$_m$) and 1350 °C (0.71 T$_m$). Mitra et al. (2003). With kind permission of Springer

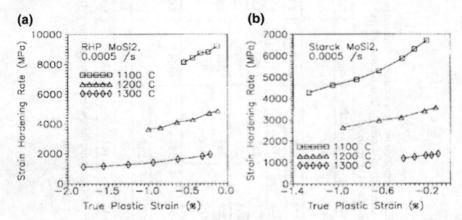

Fig. 4.30 Variation of strain hardening, dσ/dε, with true plastic strain, ε$_p$, of **a** RHP MoSi$_2$ and **b** Starck MoSi$_2$, tested at strain rates of 5×10^{-4} s^{-1}. Mitra et al. (2003). With kind permission of Springer

with fine and coarse grain sizes. In Table 4.5 the stain rate sensitivity is shown while in Tables 4.3 and 4.4 the stress exponents are listed for various strain rates. Recall that the strain hardening exponent is given by the relation of

$$\sigma = K\varepsilon^n \qquad\qquad (4.7)$$

where σ and ε are the flow stress and strain, respectively and K is a constant. The strain hardening rate is given by equation

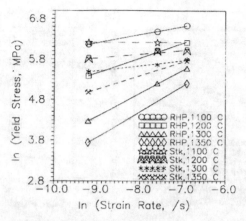

Fig. 4.31 Plots showing the variation of natural logarithms of yield stress with those of strain rates. Solid lines represent RHP MoSi$_2$, while dashed lines are for Starck MoSi$_2$. Mitra et al. (2003). With kind permission of Springer

Table 4.5 Table showing grain sizes, oxygen content, and strain rate sensitivity, m. Mitra et al. (2003). With kind permission of Springer

Materials	Mean gram size (μm)	Oxygen content (wt pct)	m at 1100 °C (0.60 T_m)	m at 1200 °C (0.64 T_m)	m at 1300 °C (0.68 T_m)	m at 1350 °C (0.71 T_m)
RHP MoSi$_2$	5.0	0.06	0.20	0.37	0.57	0.63
Starck MoSi$_2$	27.0	0.89	0.03	0.09	0.13	0.34

$$\dot{\varepsilon} = \alpha \rho_m b v \tag{4.8}$$

where a is a geometric constant dependent on the orientation of slip systems, ρ_m is the mobile dislocation density, and **b** is the Burgers vector. The variation of the strain hardening rate against the true strain is seen in Fig. 4.32.

4.3.4 WSi$_2$

(a) Single Crystal

Compression tests were performed in WSi$_2$ single crystals as a function of orientation in the temperature range of room temperature −>1500 °C. Unlike in MoSi$_2$ where plastic deformation is possible even at room temperature, WSi$_2$ can be deformed at temperatures higher than 1100 °C. Four slip systems are observed in WSi$_2$, namely

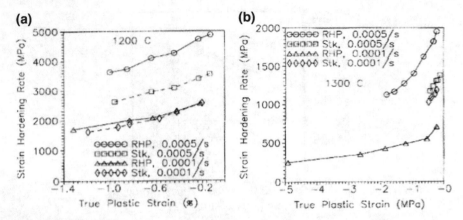

Fig. 4.32 Plots comparing the nature of variation of SHRs, dσ/dε, with true plastic strain, ε_p, of RHP MoSi$_2$ and Starck MoSi$_2$ at **a** 1200 °C and **b** 1300 °C and strain rates of 10^{-4} s^{-1} and 5 × 10^{-4} s^{-1}. Mitra et al. (2003). With kind permission of Springer

{110}⟨111⟩, {011}⟨100⟩, {023}⟨100⟩ and (001)⟨100⟩. The critical resolved shear stress values are higher in WSi$_2$ than in MoSi$_2$ as seen in Fig. 4.33.

The higher yield stress values is not only associated with the intrinsic deformation difference between MoSi$_2$ and WSi$_2$ but also because in WSi$_2$ a large number of grown-in stacking faults exist on (001) as illustrated in Fig. 4.34. The compression specimens were cut from as grown single crystals, their orientations evaluated and tested at the orientations of [0 15 1], [011], [$\bar{1}$10], [$\bar{2}$21], [$\bar{1}$12] and [$\bar{1}$13] and [001] in the [001]–[010]–[$\bar{1}$10] standard triangle. Stress-strain curves at selected temperatures for some orientations are shown in Fig. 4.35. The orientations considered in this section and seen in Figs. 4.33 and 4.35 are listed with the greatest Schmid factor for slip in Table 4.6. The yield stress variation with temperature for the orientations indicated is illustrated is seen in Fig. 4.36.

Slip lines for the orientations considered are shown in Fig. 4.37. The observed slip systems observed in WSi$_2$ which have been investigated—seven altogether— are compared in Table 4.7 with the corresponding slip systems in MoSi$_2$ above 1000 °C. The {011}⟨100⟩ slip system is operative above 1100 °C, while the {023}⟨100⟩ and {110}⟨111⟩ systems are operative above 1200 °C in WSi$_2$ single crystals. The CRSS for these three slip systems in WSi$_2$ decreases monotonically with increasing temperature without showing any anomalous increase in CRSS as indicated in Fig. 4.33.

In conclusion WSi$_2$ single crystals can be deformed only at high temperature above 1100 °C contrary to MoSi$_2$ in which deformation is possible even at room temperature. In the case of [001] orientation even a higher temperature of 1400 °C is required for plastic flow. Four slip systems are identified for slip in WSi$_2$ compared to MoSi$_2$ where only three slip systems are operative. The (001)⟨100⟩ slip is operative only in WSi$_2$ but not in MoSi$_2$ where an alternative slip system is operative, namely,

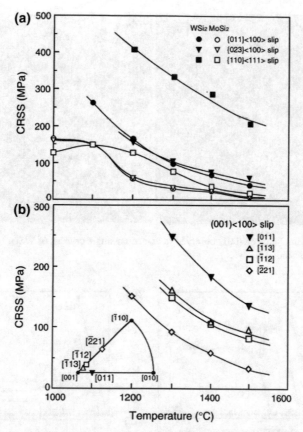

Fig. 4.33 **a** Temperature dependence of CRSS for {011}⟨100⟩, {023}⟨100⟩ and {110}⟨111⟩ slip systems obtained in WSi$_2$ and MoSi$_2$ single crystals. **b** Temperature dependence of CRSS for slip on (001)⟨100⟩ in WSi$_2$ single crystals obtained for four different orientations. Ito et al. (1999). With kind permission of Elsevier

that of {013}⟨331⟩. This system can not operate in WSi$_2$ because as indicated above a large number of stacking faults on (001) is formed during crystal growth.

4.3.5 TiSi$_2$

(a) Single Crystal

Compression experiments were performed at high temperatures on C54 TiSi$_2$ single crystals (structure oF24). The compression axes are shown in Fig. 4.38 plotted in stereographic projection. Rectangular compression specimens of about $2 \times 2 \times 5$ were used for the tests.

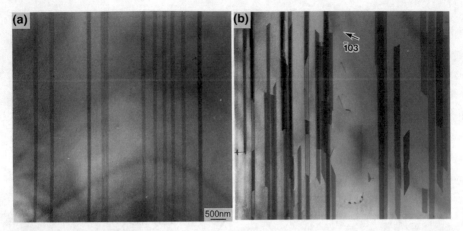

Fig. 4.34 Stacking faults on (001) observed in as-grown single crystals of WSi₂. Ito et al. (1999). With kind permission of Elsevier

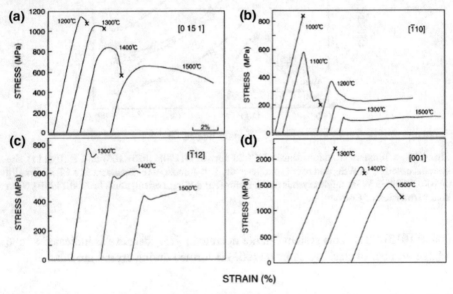

Fig. 4.35 Stress-strain curves of WSi₂ single crystals with orientations **a** [0 15 1], **b** [Ī10], **c** [Ī12] and **d** [001] at selected temperatures. Ito et al. (1999). With kind permission of Elsevier

The activation volume, ν^*, and the enthalpy, ΔH, were calculated according to the relations, respectively given as

$$\nu* = \frac{k_B T}{\phi}\left[\frac{\partial ln\dot{\varepsilon}}{\partial \sigma}\right]_T \tag{4.9}$$

Table 4.6 The largest Schmid factors for the possible slip systems for the seven different orientations investigated. Ito et al. (1999). With kind permission of Elsevier

Slip systems	Orientations						
	[0 15 1]	[011]	[$\bar{1}$10]	[$\bar{2}$21]	[$\bar{1}$12]	[$\bar{1}$13]	[001]
{013}⟨331⟩	0.392	0.498	0.387	0.277	0.447	0.443	0.387
{110}⟨111⟩	0.341	0.250	0	0.429	0.231	0.161	0
{011}⟨100⟩	0.060	0.132	0.463	0.397	0.107	0.066	0
{023}⟨100⟩	0.083	0.183	0.426	0.426	0.131	0.084	0
{001}⟨100⟩	0.159	0.350	0	0.350	0.188	0.131	0

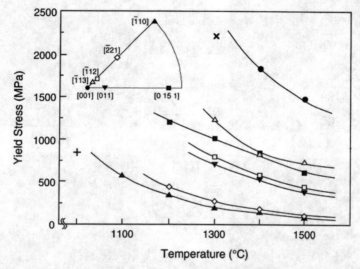

Fig. 4.36 Temperature dependence of yield stress for WSi$_2$ single crystals with orientations [001], [$\bar{1}$13], [$\bar{1}$12], [$\bar{2}$21], [$\bar{1}$10], [011] and [0 15 1]. The stresses at which fracture occurred for [001] and [$\bar{1}$10] orientations are indicated by x and +, respectively. Ito et al. (1999). With kind permission of Elsevier

$$\Delta H = -k_B T^2 \left[\frac{\partial ln\dot{\varepsilon}}{\partial \sigma} \right]_T \left[\frac{\partial \sigma}{\partial T} \right]_{\dot{\varepsilon}} \qquad (4.10)$$

Clearly $\dot{\varepsilon}$ is the strain rate, ϕ is the Schmid factor for the slip system and k_B is the Boltzmann factor. The temperature of the load versus strain for the orientations indicated in Fig. 4.38 is shown in Fig. 4.39. Following about 2% of compressive deformation the flow curves at all temperatures showed stress relaxation. In each specimen, the flow stress is lower at the higher temperatures. A definite orientation dependence is indicated in Fig. 4.39. However in all orientations the stress-strain curves exhibited a yield drop and with decreasing temperature the stress relaxation increased. In orientation c yielding could be observed up to a lower temperature without fracture. In orientation ac the specimen could be deformed up to room temperature

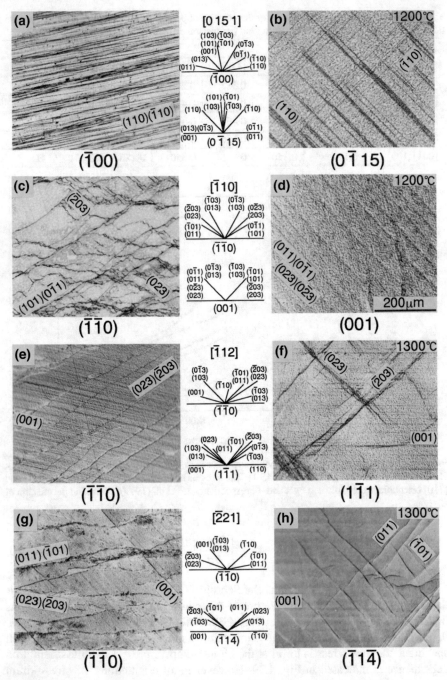

Fig. 4.37 Slip lines observed on WSi$_2$ single crystals with orientations **a** [0 15 1] at 1200 °C, **b** [$\bar{1}$10] at 1200 °C, **c** [$\bar{1}$12] at 1300 °C, **d** [$\bar{2}$21] at 1300 °C. The compression axis is in the vertical direction and traces of possible slip planes are indicated in the middle of each set of figures. Ito et al. (1999). With kind permission of Elsevier

Table 4.7 The observed slip systems in WSi$_2$ single crystals for the seven orientations investigated. Those observed for the corresponding orientations in MoSi$_2$ single crystals above 1000 °C are also indicated for comparison. Ito et al. (1999). With kind permission of Elsevier

Orientations	WSi$_2$	MoSi$_2$
[0 15 1]	{110}⟨111⟩	{110}⟨111⟩
	(001)⟨100⟩	
[$\bar{1}$10]	{011}⟨100⟩	{011}⟨100⟩
	{023}⟨100⟩	{023}⟨100⟩
[$\bar{2}$21]	(001)⟨100⟩	{110}⟨111⟩
	{011}⟨100⟩	{011}⟨100⟩
	{023}⟨100⟩	{023}⟨100⟩
[$\bar{1}$12]	(001)⟨100⟩	
	{023}⟨100⟩	
[$\bar{1}$13]	(001)⟨100⟩	{013}⟨331⟩
[011]	(001)⟨100⟩	
[001]	(001)⟨100⟩	{013}⟨331⟩

Fig. 4.38 Stereographic plot of compression axes a, b, c, ac and bc. Takeuchi et al. (1994). With kind permission of Elsevier

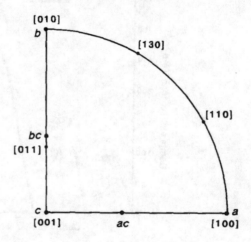

while in orientation bc serrated flow was observed in the intermediate temperature range.

The temperature dependence of the yield stress and the strain rate sensitivity (Eq. 4.5) for the orientations considered is illustrated in Fig. 4.40. The slip systems were determined to be on the (001) slip plane for orientations ac and bc and the slip lines are indicated in Fig. 4.41. When the deformation was above 1100 K the slip lines could not be resolved by optical microscopy probably due to contamination and thus the slip plane for orientations a, b and c could not be determined. The specimen could not be plastically deformed below 1000 K due to brittle fracture before yielding.

The critical resolved shear stress (CRSS) for three slip systems, (001)[110], (0$\bar{1}$1)[011] and ($\bar{3}$10)[130] were calculated. The temperature dependences of the CRSS for these three slip systems for the shear strain-rate of ~1 × 10^{-4} s^{-1} are

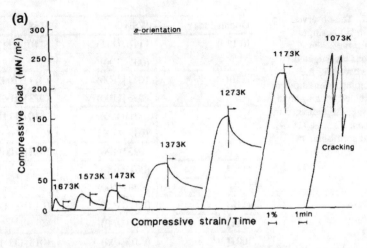

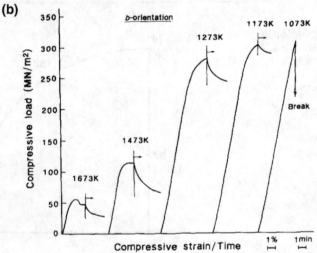

Fig. 4.39 Examples of temperature change test for specimens **a** a-orientation, **b** b-orientation, **c** c-orientation, **d** ac-orientation and **e** bc-orientation. Each curve consists of a yielding part followed by a stress-relaxation part of which the region is indicated by an arrow. Temperature has been increased or decreased. Takeuchi et al. (1994). With kind permission of Elsevier

given in Fig. 4.42. The slip system in ac-orientation taking the Schmid factors into consideration, is determined to be (001)[110] and (001) [$\bar{1}$10]. In bc-orientation, the Schmid factor for the (001)[110] and (001) [$\bar{1}$10] slip systems is about half of the Schmid factor for the same slip systems in ac-orientation. Therefore the yield stress for bc-orientation by the (001)[110] or (001) [$\bar{1}$10] slip should be about twice that for the ac-orientation. (Although the actual yield stresses is considerably higher than the values expected from the results.) It seems reasonable to conclude that the activated slip systems in bc-orientation are the same as those in ac-orientation. In

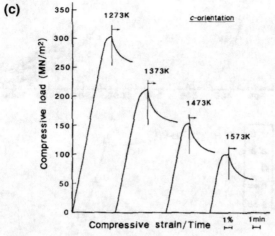

(c)

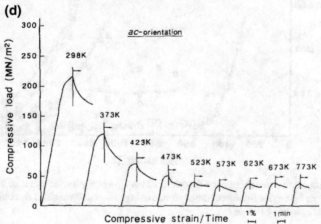

(d)

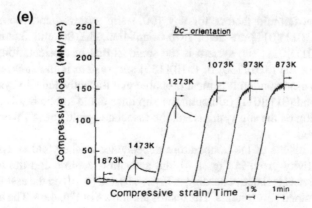

(e)

Fig. 4.39 (continued)

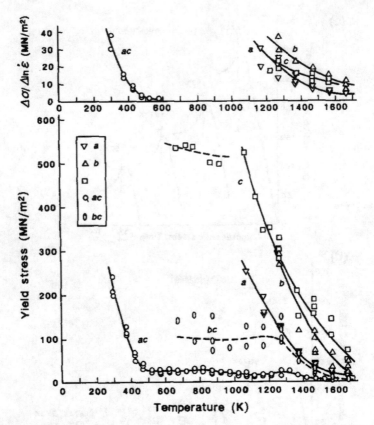

Fig. 4.40 Temperature dependence of the yield stress (lower figure) and that of the strain-rate sensitivity (upper figure) for five orientations plotted using different symbols. Takeuchi et al. (1994). With kind permission of Elsevier

a-orientation, the Schmid factors for any (001) slip systems and those for the (011)[011] and (01$\bar{1}$)[011] are zero. In b-orientation, the Schmid factor for the (011)[011] or (01$\bar{1}$)[011] slip system is the same as that in c-orientation, and the Schmid factor for the ($\bar{3}$10)[130] or (310)[1$\bar{3}$0] slip system is the same as that in a-orientation. In c-orientation it is most probable that the activated slip systems are the (011)[01$\bar{1}$] and (01$\bar{1}$)[011] (although the slip lines could not be resolved). For a detailed discussion on the slip systems and the associated yield stress Takeuchi et al. could be consulted.

Based on the results of the temperature dependence of the yield stress and the strain-rate sensitivity given in Fig. 4.40, the activation volume and the activation enthalpy have been calculated according to Eqs. (4.9) and (4.10) on the assumption of the above-determined slip systems. The results are shown in Fig. 4.43. The activation enthalpy follows an Arrhenius type relation and the analysis provide a value of ~30 for the exponent $\Delta H/k_B T$ for the four orientations seen in Fig. 4.43.

(a)

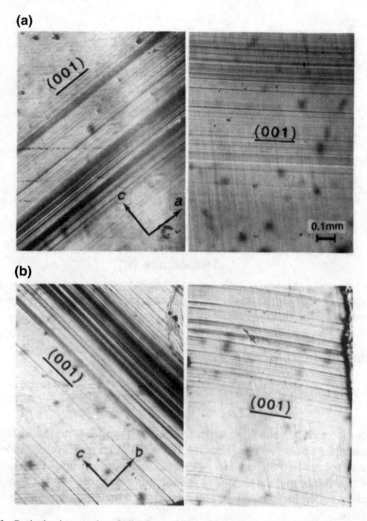

(b)

Fig. 4.41 Optical micrographs of slip lines observed on two surfaces of specimens with **a** ac-orientation and **b** bc-orientation. Deformation temperature was 673 K for ae-orientation and 973 K for bc-orientation. The compression axis is vertical. Takeuchi et al. (1994). With kind permission of Elsevier

The activation volume is 1 nm^3 which is about 10b^3 (where b = total Burgers vector) whereas in Fig. 4.43 the value at high stress is less than this value. Further the observed very large temperature dependence of the yield stress (Fig. 4.40) indicates that the deformation is controlled by the Peierls mechanism.

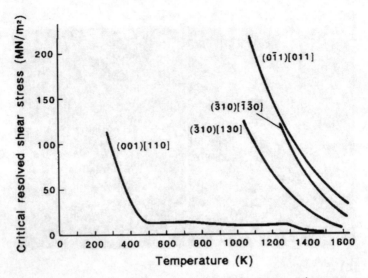

Fig. 4.42 Temperature dependence of the critical shear stress at $\dot{\gamma} \cong 10\%$ s^{-1} for three slip systems. For the (310)[130] system, CRSS depends on the sense of the shear. Takeuchi et al. (1994). With kind permission of Elsevier

4.4   Indentation—Hardness

4.4.1   Introduction

In a hardness test an indenter is pressed into a polished surface of the test material with a known load for a predetermined (specified) dwell time and the resulting indentation is measured using a microscope. The type of the indenter determines the load and the shape of indentation and the method of the measured parameter which is dependent on the geometry of the indentation. The formula of a hardness measurement is thus expressed as

$$H = \frac{P}{\text{impression}} = \frac{\text{kgf}}{\text{mm}^2} \tag{4.11}$$

H is the hardness and a letter identifying the type of indenter is added (for example for Knoop Hardness measurement the letter added is K, such as HK); P is the load applied, usually expressed in kgf; the impression can be expressed as area evaluated from the geometry of the indentation. The advantages of the test are that only a very small sample of material is required, and that it is valid for a wide range of test forces. The main disadvantages are the difficulty of using a microscope to measure the indentation (with a desirable accuracy), and the time needed to prepare the sample and apply the indenter. Variables such as load, temperature, and environment influence the results.

Fig. 4.43 Stress dependence of the activation volume and activation enthalpy analyzed according to Eqs. (4.9) and (4.10) for three slip systems. Takeuchi et al. (1994). With kind permission of Elsevier

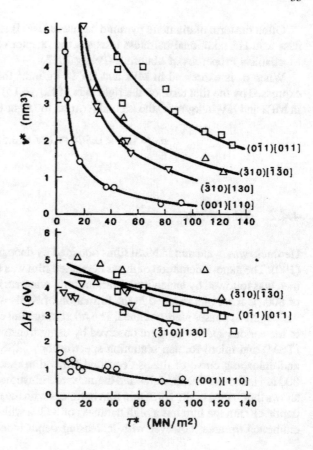

An important method of measuring hardness is the Vickers hardness, HV. Again as indicated in Eq. (4.11) the HV number is determined by force/area. The area in Vickers indentation is expressed as

$$A = \frac{d^2}{2\sin(136°/2)} \qquad (4.12)$$

This equation can be approximated by (see Pelleg 2013)

$$A \approx \frac{d^2}{1.8544} \qquad (4.13)$$

where d is the average length of the diagonal of the indentation. Thus HV can be expressed as

$$HV = \frac{P}{A} = \frac{1.8544P}{d^2} \left(\frac{kgf}{mm^2}\right) \qquad (4.14)$$

Often the term of diamond pyramid hardness (DPH) is used for the Vickers hardness test. For additional hardness tests see the chapter on hardness in the book on Mechanical Properties of Materials (Pelleg 2013).

When σ_o is expressed in MPa and HV in kg/mm^2 the 0.2% yield stress can be expressed by the first term on the right side of Eq. (4.15), and when σ_o is expressed in MPa and HV in kg/mm^2 the second term on the right is valid.

$$\sigma_0 \approx HV \times 0.364 \approx HV \times 3.55 \tag{4.15}$$

4.4.2 NiSi$_2$ Film

Hardness was evaluated in Ni/Si films obtained by depositing 100 nm layer Ni on Si (100). The nano-indentation of the as deposited film was to a depth of 500 nm which then was followed by annealing at various temperatures. Annealing at a temperature of 800 °C for 2 min resulted in the formation of NiSi$_2$ while at lower temperatures various Ni/Si phases were obtained. The microstructural changes and phases induced in the various specimens were observed by using transmission electron microscopy (TEM) and micro-Raman scattering spectroscopy (RSS). In Fig. 4.44 the loading and unloading curve of the as deposited Ni/Si film indented to depths of 500 and 800 nm is shown. The modulus and hardness are illustrated in Fig. 4.45. Figure 4.44b shows the change of the Ni/Si thin film hardness with the nanoindentation depth. At a depth <10 nm the film has a high hardness of 8 GPa which is obtained due to poorly calibrated tip area, however with increasing depth indentation the hardness drops

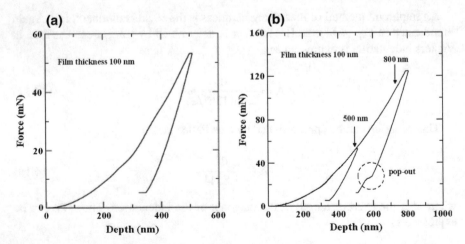

Fig. 4.44 Loading–unloading curve for as-deposited Ni/Si thin film indented to a maximum depth of **a** 500 nm; **b** comparison of loading–unloading curve of as-deposited Ni/Si thin film indented to different depth of 500 and 800 nm. Lee et al. (2010). With kind permission of Springer

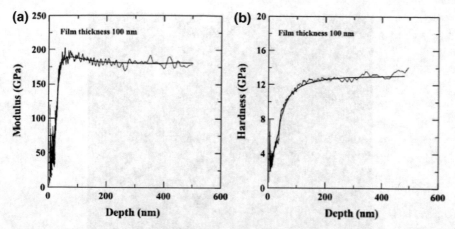

Fig. 4.45 **a** Variation of Young's modulus of Ni/Si thin film with indentation depth; **b** variation of hardness of Ni/Si thin film with indentation depth. Lee et al. (2010). With kind permission of Springer

rapidly to 3.5 GPa at an indentation depth of 15 nm. But then the hardness increases gradually as the indenter penetrates deeper to the silicon substrate.

At the final indentation depth of 500 nm the hardness has a value of approximately 13 GPa, which is the same as that of the Si (100) substrate (13 GPa). The hardness of the Ni/Si thin-film system can be estimated as

$$H_{eff} = H_s + (H_f - H_s)\varphi_H \tag{4.16}$$

H_s and H_f are the hardness of the substrate and the film, respectively, and φ_H is a weighting function. As shown in Fig. 4.45b, the hardness of the Ni/Si thin film increases rapidly as the indentation depth increases.

Clearly other properties change with the indentation. For example the Young's modulus change with indentation in a similar fashion to the hardness as seen in Fig. 4.45a. Following the dip to 8 GPa (for the reason mentioned above) at an indentation <10 nm with further indentation depth the Young modulus increases, reaching a value of 188 GPa at an indentation of about 100 nm. This is a consequence of the strain gradient hardening effect. At greater indentation depth the indenter tip penetrates the substrate and therefore the modulus decreases to about 180 GPa. When the indenter completely penetrates into the silicon substrate the modulus reaches the constant value of the silicon substrate at 177 GPa. The relation has the same form as that of the hardness namely:

$$E_{eff} = E_s + (E_f - E_s)I_0 \tag{4.17}$$

E_f and E_s are the film and silicon moduli respectively and I_0 is a function of t/a where t is the film thickness, and, a being the constant radius.

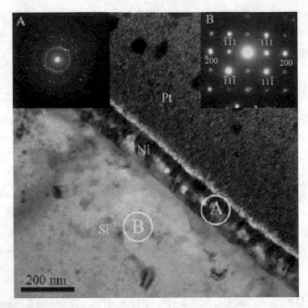

Fig. 4.46 Bright field TEM micrograph of deposited Ni/Si thin film. Lee et al. (2010). With kind permission of Springer

TEM micrograph of the as deposited film and the one annealed at 800 °C for 2 min are compared in Figs. 4.46, 4.47 and 4.48. At this temperature the film completely transformed to the NiSi$_2$ phase. In the selected area diffraction pattern of Fig. 4.48 the inset A contains in addition to the diffraction pattern Si spots also, which indicates that the microstructure contains only epitaxial NiSi$_2$. The NiSi$_2$ phase at 800 °C was confirmed by Raman spectrum. The lower temperature annealed specimens indicate that the NiSi phase disappears at about 750–800 °C.

The phases developed in the system depend on the annealing temperature. For observing the microstructures formed at temperatures lower than 800 °C the reader can consult the works of Lee et al. (2010).

4.4.3 MoSi$_2$ Single Crystal

Within the framework of deformation experiments, the mechanical properties of single crystal MoSi$_2$ were investigated among them the Vickers hardness was included in order to evaluate its structural applications. The MoSi$_2$ single crystals were grown from a master ingot by floating zone technique. Micro-Vickers hardness (at three different orientations) were evaluated as a function of temperature as shown in Fig. 4.49.

Three regions can be observed in the curves: first a sharp drop in the hardness values occur with increasing temperature; then a plateau-like region possibly with gradual hardness decrease is seen and in the third region again a fast drop in the

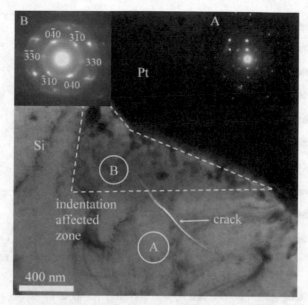

Fig. 4.47 Bright field TEM micrograph as-of deposited indented specimen. Lee et al. (2010). With kind permission of Springer

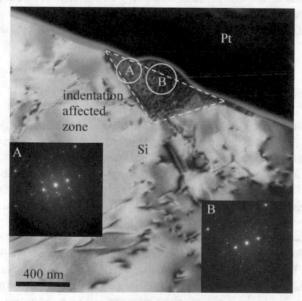

Fig. 4.48 Bright field TEM micrograph of indented specimen annealed at 800 °C for 2 min (diffraction patterns of zones A and B shown in insets). Lee et al. (2010). With kind permission of Springer

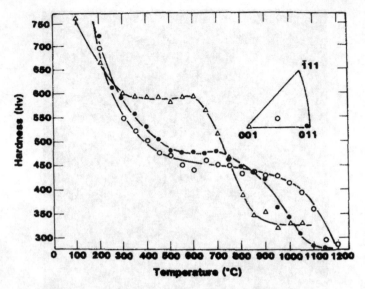

Fig. 4.49 The temperature dependence of micro-Vickers hardness of MoSi$_2$ single crystals. Umakoshi et al. (1989). With kind permission of Elsevier

hardness values occurs. The indenter with 300 g was applied for 10 s. The hardness values are orientation dependent. In the first decrease in the hardness slip traces are observed around the indentation along {110} and microcracks also develop along the {010}.

4.4.3.1   MoSi$_2$ Film

MoSi$_2$ film was produced by magnetron sputter deposition from a hot pressed MoSi$_2$ target. The as-deposited film on silicon (111) substrate was amorphous and therefore crystallization was induced by annealing. The interest in MoSi$_2$ is due to its attractive properties. It is a metal-like stable conductor, has exceptional oxidation resistance in air up to ~1700 °C and therefore used among its other applications as electrical heating element in high temperature furnaces. To obtain the tetragonal stable MoSi$_2$ structure high temperature annealing of ~900 °C or above is required. At lower temperatures,—say annealing at ~800 °C—resulted in hexagonal MoSi$_2$ from the amorphous MoSi$_2$ film. The depth concentration profile of the as-deposited MoSi$_2$ film obtained by Auger electron spectroscopy (AES) is illustrated in Fig. 4.50. In Fig. 4.51 TEM images are illustrated. The hardness of the MoSi$_2$ film is shown in Fig. 4.52. Included in the figure are the hardnesses of the hot pressed MoSi$_2$ target, the silicon (111) surface and the variation of the fused quartz which was used for calibration purposes (to determine the shape function the diamond indenter). It can be seen that there is no difference in the hardness values between the MoSi$_2$ bulk and

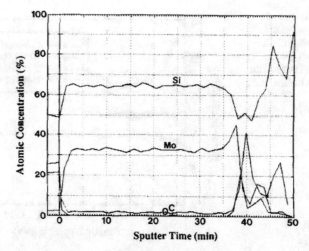

Fig. 4.50 AES depth concentration profiles of an as-deposited MoSi$_2$ film. Chou and Nieh (1992). With kind permission of Elsevier

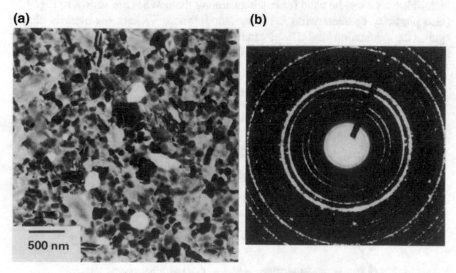

Fig. 4.51 a A TEM bright field image and **b** the corresponding electron diffraction pattern of an MoSi, film after annealing at 900 °C 2 h. Chou and Nieh (1992). With kind permission of Elsevier

film, thus an indication is obtained about the hardness expectation in bulk MoSi$_2$. The hardness is quite constant with penetration depth (plastic region).

Fig. 4.52 Hardnesses of
hot-pressed MoSi₂ bulk,
as-deposited MoSi₂ films,
Si(111) and fused quartz.
Chou and Nieh (1992). With
kind permission of Elsevier

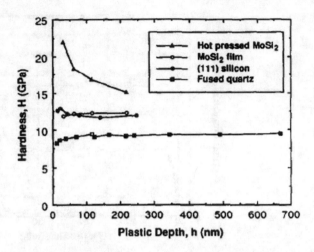

4.4.4 WSi₂ Film (Coating)

Coating is one of the applications of WSi₂ film among its other uses. Some typical materials which can be used for coatings among them WSi₂ are shown in Fig. 4.53 (also plasticity characteristics δH is included) where Vickers hardness is plotted against the indentation load. The plot indicates a small (almost insignificant) decrease in hardness with increasing indentation load (at the same time a small increase in

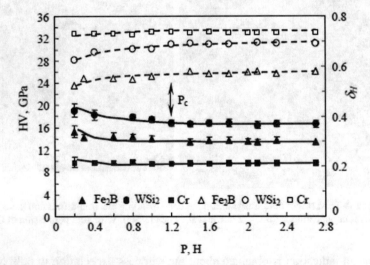

Fig. 4.53 A summary of the data for mechanical parameters of coatings versus indentation load P. Closed and open symbols denote Vickers hardness, HV, and plasticity characteristic δH (open symbols), respectively. Byakova et al. (2004). With permission of Dr. Nina Obradovic for Sci. of Sintering Editorial Board

δH occurs also insignificant) until its stabilization. The final values at each indentation load was averaged over 10 indentations. The hardness value of WSi_2 is HV = 5.63 GPa.

4.4.5 α-FeSi₂

Important parameters for designing and developing new alloys are the properties of hardness, fracture toughness and Young's moduli. K_{IC} is an important measure of toughness because it is independent of specimen geometry. Empirical methods exist to calculate the fracture toughness, K_{IC} for a given material from Vickers microhardness measurements (Palmqvist 1962) but later other relations were also used as indicated in Table 4.8.

The hardness measurement were performed with three loads of 0.3, 0.5 and 1 kg. Ten indentations were made for each load thus the results provide reasonable statistics. Cracks formation were observed as seen in Fig. 4.54 indicating also the microhardness indentation. The cracks that had developed from the indentation marks were of the Palmqvist type. The longest crack length from the centre of the indentation mark to the crack tip was measured for each indentation. The hardness value obtained for the investigated α-FeSi₂ is 5.63 GPa. The direct measurements of the hardness were intended to evaluate from these data the fracture toughness K_{IC} according the equations indicated in Table 4.6. Also, the plots shown in Fig. 4.55 with coordinates x and y are explained in the relations of Table 4.8. The values of fracture toughness, K_{IC} were determined from three-point bending tests and were compared with values calculated from Vickers indentations. The required elastic modulus for the calculations according to relations 1–4 of Table 4.8 were determined by compression

Table 4.8 Proposed relationships between K_{IC}, E and data obtained from Vickers hardness measurements. Milekhine et al. (2002). With kind permission of Elsevier

Eq.	Author	Equation	Rearranged equation $y = \text{const.} \times x$
(1)	Niihara et al. (1982)	$K_{IC} = 0.0089 \cdot \left(\frac{E}{H_v}\right)^{2/5} \cdot \frac{P}{a \cdot l^{1/2}}$ for $2.5 \geq l/a \geq 0.25$	$a \cdot l^{1/2} =$ $E^{2/5}/K_{1C} \cdot 0.0089 \cdot \frac{P}{H_v^{2/5}}$
(2)	Niihara (1983)	$K_{IC} = 0.0122 \cdot \left(\frac{E}{H_v}\right)^{2/5} \cdot \frac{P}{a \cdot l^{1/2}}$ for $2.5 \geq l/a \geq 1$	$a \cdot l^{1/2} =$ $E^{2/5}/K_{1C} \cdot 0.0122 \cdot \frac{P}{H_v^{2/5}}$
(3)	Shetty et al. (1985)	$K_{IC} = 0.0319 \cdot \frac{P}{a \cdot l^{1/2}}$	$a \cdot l^{1/2} = \frac{1}{K_{IC}} \cdot 0.0319 \cdot P$
(4)	Laugier (1987)	$K_{IC} =$ $0.0143 \cdot \left(\frac{E}{H_v}\right)^{2/3} \cdot \left(\frac{a}{l}\right)^{1/2} \frac{P}{c^{3/2}}$	$\left(\frac{1}{a}\right)^{1/2} c^{3/2} =$ $E^{2/3}/K_{IC} \cdot 0.0143 \cdot \frac{P}{H_v^{2/3}}$

H_v—Vickers hardness number, P—load during Vickers test, $2a$—indentation mark diagonal, and $l = c - a$ where c is total crack length from centre of hardness mark

Fig. 4.54 Microcracks from
the corners of a hardness
indentation mark. Milekhine
et al. (2002). With kind
permission of Elsevier

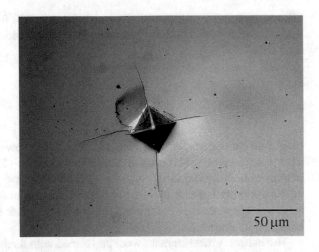

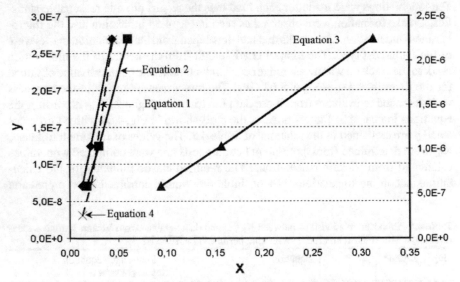

Fig. 4.55 Graphic representation of the experimental microhardness data for the α-FeSi$_2$ phase.
The symbols indicate the equations of Table 4.8 used for the explanation of x and y. Thus for Eq. (1)
diamonds, for Eq. (2) rectangles, for Eq. (3) triangles for Eq. (4) dashed line and crosses are used.
Milekhine et al. (2002). With kind permission of Elsevier

tests. For this section the hardness value are of interest, and it is of no relevance if
the original tendency was to show that Vickers hardness measurements can be used
for fracture toughness evaluation, an experiment which is simpler, cost saving and
requiring only simple hardness measuring device. In the relations of Table 4.8, the
values of the half indentation, a, and the crack length from the centre of indentation,
c are required. Thus, l, which is also required in the equations is given by $l = c - a$.
These parameters are obtained from the Vickers hardness measurements.

4.4.6 TiSi₂

In a relatively recent work the properties of TiSi$_2$ were measured, including its Vickers hardness. The TiSi$_2$ was synthesized and consolidated from Ti and Si powders by pulsed current activated combustion technique within 1 min in a one step process. The consolidation was performed by a combined effect of combustion synthesis and the application of a mechanical pressure. Under the simultaneous application of a pulsed current and the 60 MPa pressure a density of 96% was achieved. The starting constituents is illustrated in Fig. 4.56. The variation of temperature with dwell time of the sample to obtain a dense consolidated TiSi$_2$ is illustrated in Fig. 4.57.

Further, the images at the various stages of the processing are indicated in Fig. 4.58. The Vickers hardness was performed on polished TiSi$_2$ using a load of 10 kgf at a dwell time of 15 s, resulting in a hardness of 964 kg/mm^2. The value is the average of five measurements. At some larger load crack developed around the indenter. The indentation pattern and a crack formed during the indentation are illustrated in Fig. 4.59. The value of the hardness and the measured trace length of the crack the fracture toughness, K$_{IC}$ could be estimated according to the relation (Anstis et al. 1981).

$$K_{IC} = 0.016(E/H)^{1/2}(P/C)^{3/2} \qquad (4.18)$$

Also the toughness value is an average of five measurements. Typically, 1–3 cracks were observed to propagate from the indentation corner which seems to propagate linearly. Clearly in the relation H is hardness, E the Young's modulus for TiSi$_2$ is 255.6 GPa, P is the applied load and C is the trace length of the crack. K$_{IC}$ ~ 2.5 MPa/m^2. The Vickers hardness value for TiSi$_2$ reported by Frommeyer and Rosenkranz (2004) in an earlier report in his Table 1 is 870 ± 15HV.

Fig. 4.56 Scanning electron microscope images of the raw materials: **a** titanium, **b** silicon powder. Kim et al. (2009). With kind permission of Dr. Kim for the authors

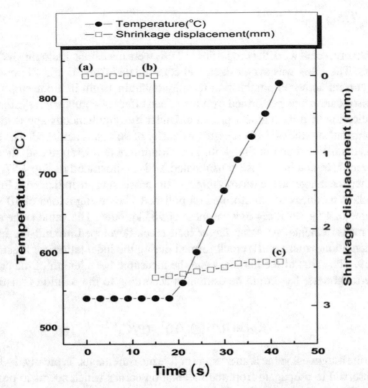

Fig. 4.57 Variations of temperature and shrinkage displacement with heating time during high-frequency induction heated combustion synthesis and densification of TiSi$_2$ (under 60 MPa, 90% output of total power capacity). Kim et al. (2009). With kind permission of Dr. Kim for the authors

Fig. 4.58 Scanning electron microscope images of Ti + Si system: **a** after milling, **b** before combustion synthesis, **c** after combustion synthesis. Kim et al. (2009). With kind permission of Dr. Kim for the authors

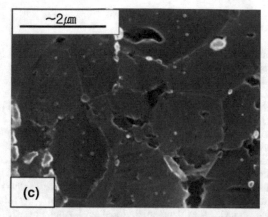

Fig. 4.58 (continued)

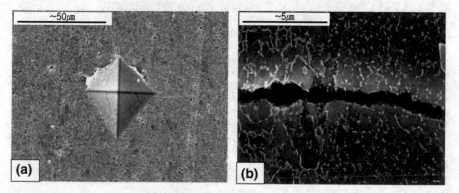

Fig. 4.59 **a** Vickers hardness indentation and **b** median crack propagating of $TiSi_2$. Kim et al. (2009). With kind permission of Dr. Kim for the authors

Summary

- Tests are performed by tension, compression and indentation (hardness)
- Single and/or polycrystalline silicides ($CoSi_2$, $NiSi_2$, $MoSi_2$, WSi_2 and $TiSi_2$) were deformed by tension. Some of them were thin films
- Whenever available the orientation dependent deformation at various temperatures are considered
- Compression tests of some of the silicides indicate the effect of the grain size and orientation
- Hardness tests mainly by Vickers indentation of most of the silicides show the possibility of fracture toughness estimation by hardness measurements
- Relations between hardness and tensile properties (yield stress), and other important equations related to stress evolution in silicides are included. Other relevant relations for deformation are indicated
- Crack formation during indentation can be avoided by proper load application.

References

G.R. Anstis, P. Chantikul, B.R. Lawn, D.B. Marshall, J. Am. Ceram. Soc. **64**, 533 (1981)
W.A. Brantley, J. Appl. Phys. **44**, 534 (1973)
P. Burkhardt, R.E. Marvel, J. Electrochem. Soc. **116**, 864 (1969)
A.V. Byakova, Yu.V. Milman, A.A. Vlasov, Sci. Sinter. **36**, 93 (2004)
H.-Y. Chen, C.-Y. Lin, C.-C. Huang, C.-H. Chien, Microelectron. Eng. **87**, 2540 (2010)
T.C. Chou, T.G. Nieh, Thin Solid Film **214**, 48 (1992)
G. Frommeyer, R. Rosenkranz, in *Metallic Materials with High Structural Efficiency*, ed. by O.N.
    Senkov et al. (Kluwer Academic Publishers, the Netherlands, 2004), p. 287
S. Guder, M. Bartsch, M. Yamaguchi, U. Messerschmidt, Mater. Sci. Eng. A **261**, 139 (1999)
K. Ito, H. Inui, T. Hirano, M. Yamaguchi, Mater. Sci. Eng. A **152**, 153 (1992)
K. Ito, H. Inui, T. Hirano, M. Yamaguchi, Acta Metall. Mater. **42**, 1261 (1994)
K. Ito, T. Yano, T. Nakamoto, H. Inui, M. Yamaguchi, Acta Mater. **47**, 837 (1999)
J.F. Jongste, O.B. Loopstra, G.C.A.M. Janssen, S. Radelaar, J. Appl. Phys. **73**, 2816 (1993)
L. Junker, M. Bartsch, U. Messerschmidt, Mater. Sci. Eng. A **328**, 181 (2002)
M.T. Laugier, J. Mater. Sci. Lett. **6**, 355 (1987)
W.-S. Lee, T.-H. Chen, C.-F. Lin, J.-M. Chen, Appl. Phys. A **100**, 1089 (2010)
J.F. Liu, J.Y. Feng, W.Z. Li, Nucl. Instr. Methods Phys. Res. B **194**, 289 (2002)
B.-R. Kim, K.-Seok Nam, J.-K. Yoon, J.-M. Dohc, K.-T. Lee, I.-J. Shon, J. Ceram. Process. Res.
    10, 171–175 (2009)
V. Milekhine, M.I. Onsoien, J.K. Solberg, T. Skaland, Intermetallics **10**, 743 (2002)
R. Mitra, N.E. Prasad, S. Kumari, A.V. Rao, Met. Mater. Trans. **34A**, 2003 (2003)
S.P. Murarka, D.B. Fraser, J. Appl. Phys. **51**, 342 (1980)
T. Nakano, M. Azuma, Y. Umakoshi, Acta Mater. **50**, 3731 (2002)
K. Niihara, J. Mater. Sci. Lett. **2**, 221 (1983)
K. Niihara, R. Morena, D.P.H. Hasselman, J. Mater. Sci. Lett. **1**, 13 (1982)
S. Palmqvist, Arch. Eisenhuttenwesen. **9**, 1 (1962)
J. Pelleg, *Mechanical Properties of Materials* (Springer, Dordrecht, 2013)
J. Pelleg, Mechanical Properties of Ceramics (Springer, 2014)
T.F. Retajzxyk Jr., A.K. Sinha, Thin Solid Films **70**, 241 (1980)
D. Sander, A. Enders, J. Kirschner, Appl. Phys. Lett. **67**, 1833 (1995)
D.K. Shetty, I.G. Wright, P.N. Mincer, A.H. Clauer, J. Mater. Sci. **20**, 1873 (1985)
S. Takeuchi, T. Hashimoto, M. Nakamura, Intermetallics **2**, 289 (1994)
Y. Umakoshi, T. Hirano, T. Sakagami, T. Yamane, Scr. Met. **23**, 87 (1989)
Y. Umakoshi, T. Sakagami, T. Hirano, T. Yamane, Acta Metall. Mater. **38**, 309 (1990)
A.H. van Ommen, C.W.T. Bulle-Lieuwma, C. Langereis, J. Appl. Phys. **64**, 2708 (1988)
J.B. Wachtman Jr., W.E. Tefft, D.G. Lam Jr., R.P. Stinchfield, J. Res. Natl. Bur. Std. Sect. A **64**, 213
    (1960)
J.B. Wachtman Jr., T.G. Scuderi, G.W. Cleek, J. Am. Ceram. Soc. **45**, 319 (1962)

Chapter 5
Dislocations in Silicides

Abstract Dislocations in epitaxial thin films and single crystals are considered in details in the $CoSi_2$, $NiSi_2$, $MoSi_2$, WSi_2 and $TiSi_2$ silicides. Epitaxial single crystals can be obtained in ultra high vacuum by reacting a metal for example Co with a substrate Si (111) or $NiSi_2$ formation on low index planes. Often—depending on conditions—dislocations dissociate into partials forming a stacking fault with the partial dislocations as indicated for $MoSi_2$, WSi_2 and $TiSi_2$. A concept for dissociation (for example in $MoSi_2$) is that dislocations move into a nonplanar configuration by a combination of glide and climb. The dislocation structure is orientation and temperature dependent and at high temperatures their structure might be irregular. Also, at the higher temperatures dislocation-vacancy interaction may occur resulting in restrictions in the dislocation line. Serrated flow stress is believed to be the result of dislocation-vacancy reaction. At the temperature where the stress-strain curves showed serrations, the dislocations observed are wavy, whereas in many of the structures ($MoSi_2$, WSi_2) the dislocations are straight.

5.1 Introduction

The mechanical properties of materials are determined by line defects known as dislocations the existence of which was postulated originally and independently by Taylor (1934), Orowan (1934) and Polanyi (1934). They have explained the difference in the theoretical and experimental strength of material and their theory sheds light on many aspects of the mechanical properties of solids and their growth. Their postulate for solids has been confirmed by TEM, field ion microscopy and atom probe techniques directly and by etch-pit for example indirectly. Dislocations has been discussed in separate chapters in the books of Pelleg (2013, 2014) and for extending their scope of interest in the subject the interested readers are referred to them.

© Springer Nature Switzerland AG 2019
J. Pelleg, *Mechanical Properties of Silicon Based Compounds: Silicides*,
Engineering Materials, https://doi.org/10.1007/978-3-030-22598-8_5

5.2 Dislocations in CoSi$_2$

5.2.1 Single Crystals

In Sect. 4.3.2 compression tests were indicated for CoSi$_2$ where stress strain curves, yield stress information and slip lines were considered. The crystals were deformed by slip on (010) and (001) and all specimens for TEM (thin foils) were cut parallel to (010) slip planes. In Fig. 5.1a at 400 °C long and straight dislocations with their Burgers vector $\mathbf{b} = [001]$ are seen to lie on (010) slip planes, suggesting the operation of the Peierls mechanism. Most dislocations are of edge orientation. In Fig. 5.1b at 600 °C, smoothly curved dislocations and small dislocation loops are seen in addition to the long and straight edge dislocations with $\mathbf{b} = [001]$. It is considered that the

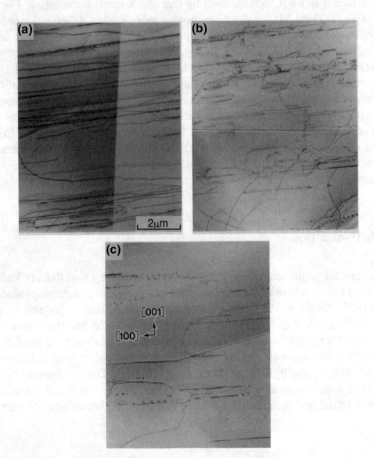

Fig. 5.1 Dislocation structures in $[\bar{1}23]$-oriented single crystals deformed at **a** 400 °C, **b** 600 °C and **c** 800 °C. All the thin foils were cut parallel to the (010) slip plane. Ito et al. (1994). With kind permission of Elsevier

Burgers vector of the small loops and the curved dislocations is $\mathbf{b} = [001]$ the same as that of the edge dislocations since they show the same contrast behavior. The dislocation density is much reduced on temperature increase as seen in Fig. 5.1c at 800 °C. A few long edge dislocations with $\mathbf{b} = [001]$, some curved dislocations and many small dislocation loops with a tendency to align along the [100] direction (i.e. perpendicular to their Burgers vector) are observed. It is probable that these loops are formed by the break-up of edge dipoles by a climb mechanism. Climb requires point defects, and thus at this temperature the deformation is influenced by point defects. At higher temperature the mobility of point defects more and more influence deformation.

Figure 5.2 shows the dislocation structure in [001] oriented single crystals deformed to about 2% strain at 700 °C. In Fig. 4.18b it was seen that the slip planes are {111} in foils cut parallel to the (111) planes. Most of the dislocations seen have long segments along ⟨110⟩ directions exhibiting many jogs. The motion of this dislocation is controlled by Peirels mechanism (this is suggested by the dislocations existing in long segments along ⟨110⟩) as in the case of {001}⟨100⟩ slip. The Burgers vectors were determined by contrast analysis and the results are shown in Fig. 5.3. Dislocations marked A have a long segment along $[\bar{1}01]$. These dislocations are in contrast in Fig. 5.3a $\mathbf{g} = \bar{2}20$ with a near [111] zone axis (ZA), in Fig. 5.3b $\mathbf{g} = 02\bar{2}$, ZA = [111], in Fig. 5.3d $\mathbf{g} = 3\bar{1}\bar{1}$, ZA $[12\bar{1}]$ and in Fig. 5.3e $\mathbf{g} = \bar{4}00$, ZA = [011] but out of contrast in Fig. 5.3c $\mathbf{g} = 1\bar{3}1$, ZA = [211], and in Fig. 5.3f $\mathbf{g} = 040$, ZA = [101].

Thus, the Burgers vector of dislocations marked A is $1/2[\bar{1}01]$. On the other hand, dislocations marked B with a long segment along $[0\bar{1}1]$ are in contrast in Fig. 5.3a–c, f but out of contrast in Fig. 5.3d, e. Thus, the Burgers vector of dislocations marked B is $1/2[0\bar{1}1]$. These dislocations are in screw orientation and believed to be trapped in Peirels valleys along their screw orientation. Thus the indications are that [001] oriented crystals slip on {111}⟨110⟩ systems at temperatures up to 700 °C.

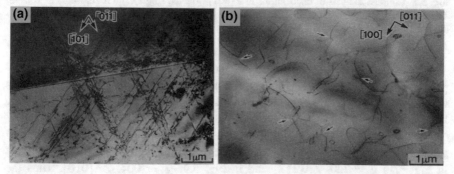

Fig. 5.2 Dislocation structures in [001]-oriented single crystals deformed at **a** 700 °C and **b** 800 °C. The thin foils were cut parallel to (111) and $(0\bar{1}1)$ slip planes for (**a**) and (**b**), respectively. Ito et al. (1994). With kind permission of Elsevier

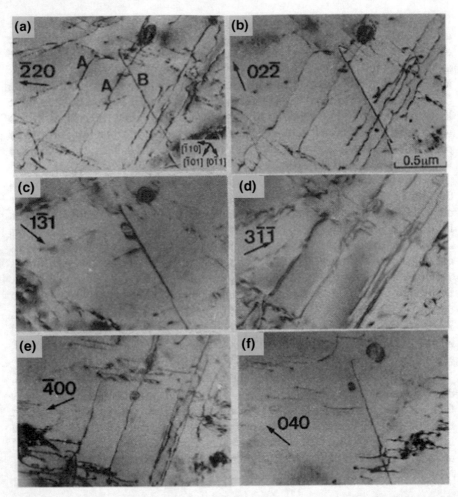

Fig. 5.3 Contrast analysis of dislocations in a [001]-oriented single crystal deformed at 700 °C. The thin foil was cut parallel to the (111) slip planes and the operating reflections are indicated in the figure. Ito et al. (1994). With kind permission of Elsevier

The dislocation density at 800 °C—as indicated earlier—greatly decreased compared to 700 °C despite the similar amount of strain applied. Again contrast analysis was performed to determine the Burgers vector of the dislocations shown in Fig. 5.2b. Details are seen in Fig. 5.4.

Dislocations A are in contrast in Fig. 5.4a $\mathbf{g} = 202$, ZA $= [\bar{1}2\bar{1}]$, in Fig. 5.4b $\mathbf{g} = 0\bar{4}0$, ZA $= [001]$ and in Fig. 5.4c $\mathbf{g} = 20\bar{2}$, ZA$[1\bar{2}0]$ but out of contrast in Fig. 5.4d $\mathbf{g} = 220$, ZA$[1\bar{1}2]$ and in Fig. 5.4e $\mathbf{g} = 00\bar{4}$, ZA$[0\bar{1}0]$. Thus the Burgers vector of the A dislocations is $1/2[\bar{1}10]$. The Burgers vectors of dislocations throughout the area imaged are $1/2\langle110\rangle$ as illustrated schematically in Fig. 5.4f. Burgers vector exist even though no shear stress is resolved in [110] and $[1\bar{1}0]$ for this

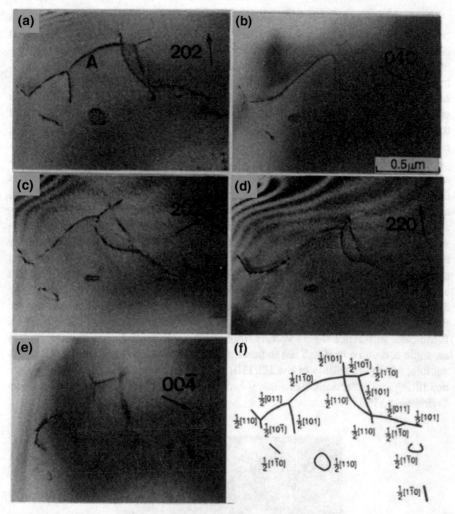

Fig. 5.4 Contrast analysis of dislocations in a [001]-oriented single crystals deformed at 800 °C. The thin foil was cut parallel to the $(0\bar{1}1)$ slip planes and the operating reflections are indicated in the figure. **f** is the schematic illustration of the results. Ito et al. (1994). With kind permission of Elsevier

orientation as is indicated in Table 4.1. This is likely to indicate that operation of slip on $\{110\}\langle110\rangle$ is necessary for CoSi₂ to be deformable by slip. As seen in Table 4.1 the Schmid factor for slip on $\{110\}\langle110\rangle$ in $[\bar{1}11]$-oriented crystals is essentially zero.

Additional details on slip were discussed in Sect. 4.2. For a comprehensive analysis of dislocations reactions related to deformation in CoSi₂ the reader is referred to the publication of Ito et al.

Dislocations and plasticity were investigated in [113] oriented crystals also between room temperature and 1173 K. Below $0.5 \, T_m = 800$ K the slip system is $\{001\}\langle010\rangle$. The glide dislocation has a Burgers vector of $\mathbf{b} = \langle010\rangle$ and is dissociated into two partial dislocations and a stacking fault between them. The partials have the same parallel Burgers vector of $1/2\langle010\rangle$. Above $0.5 \, T_m$ the mode of deformation changes to the $1/2\{001\}\langle0\bar{1}1\rangle$ systems. These systems originate from cross slip $a/2\langle110\rangle$ dislocations dissociations into two partials with Burgers vectors of the type $a/4\langle111\rangle$, $a/6\langle112\rangle$ and $a/2\langle010\rangle$ when gliding in $\{011\}$, $\{111\}$ and $\{001\}$, respectively.

At T < 800 K (regime 1) the slip systems are of the (010)(001) type. In [113] orientation there are four equally stressed cubic systems (the common Schmid factor is 0.27) namely, 1 the coplanar [010](001) and [l00](001) systems and 2 the [001](100) and [001](010) systems, which have the same slip direction. All these four systems are activated since the beginning of plastic deformation. Not all the dislocation analysis are considered here. In Fig. 5.5 bowed out $\pm a[010](001)$ dislocations with nearly a straight segment is illustrated. The observation in Fig. 5.5 strongly suggests that a lattice friction force might be the possible cause for the presence of the straight edge segments, which is stronger at the edges. A typical example of regime 2 above 800 K as observed in a (001) foil is shown in Fig. 5.6. Two dislocation c and d are present at point X and they are invisible for 111 reflection (see about X in Fig. 5.6b). It indicates different Burgers vectors from $\langle010\rangle$. Various dislocations analysis (for example c, d and e in Fig. 5.6a) indicate a common behavior regarding the Burgers vectors, which are mainly of the $1/2(110)$ type. Dislocations were analyzed at various tilting and reflections (see Figs. 9 and 10 in the original work of Anongba and Steinemann (1993)).

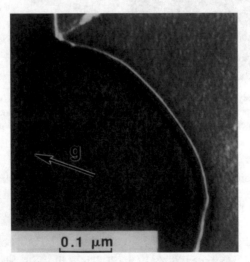

Fig. 5.5 Bowed out $\pm a[010](001)$ dislocation with a straight nearly edge segment extending over 0.3 μm in the foil. (001) foil; weak beam; $\mathbf{g} = [0\bar{2}0]$, BD∥[001]. BD is beam direction. Anongba and Steinemann (1993). With kind permission of John Wiley and Sons

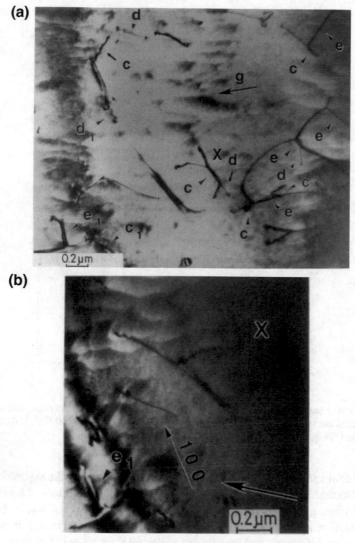

Fig. 5.6 Dislocation structure generated in [113] CoSi$_2$ single crystals by 973 K compression (regime 2). $\dot{\gamma} = 5.3 \times 10^{-5}$ s^{-1}, $\tau = 37.40$ MPa, $\gamma = 12.5\%$; (001) foil; bright field. **a** c, d and e are, respectively, $\pm a/2[0\bar{1}1]$, $\pm a/2[\bar{1}01]$ and $\pm a/2[110]$ dislocations; **g** = [040] (dislocations d display residual contrasts); BD∥[001], **b** Zone X Fig. 5.6a viewed in reflection corresponding to **g** (unlabled arrow) equal to [111]; BD∥$[0\bar{1}1]$ (same scale-mark as in Fig. 5.6a). Anongba and Steinemann (1993). With kind permission of John Wiley and Sons

It might be of interest to indicate the stress-strain curves for the slip systems associated with the dislocation motion and their dissociation. In Fig. 5.7 representative curves of regime 1 are illustrated, while those of regime 2 are shown in Fig. 5.8. The stacking fault energy, $\bar{\gamma}$ resulting from the dislocations dissociation has been

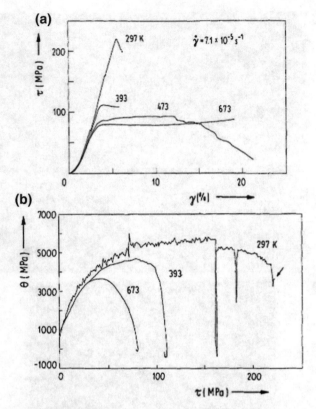

Fig. 5.7 a $\tau(\gamma)$ curves (compression test results) of [113] CoSi$_2$ single crystals and **b** three of the associated rates of hardening in temperature regime 1 for $\dot{\gamma} = 7.1 \times 10^{-5}$ s^{-1}. Anongba and Steinemann (1993). With kind permission of John Wiley and Sons

calculated for various slip systems (listed in Table 5.1) by using the separation width of the dissociated partial dislocations a/2⟨110⟩ with Burgers vectors of a/4⟨111⟩ and a/6⟨112⟩ when gliding in {011} and {111} planes, respectively. Δ_{obs} is the image peak separation and the corrected value of the partials separation is Δ. $\bar{\gamma}$ is calculated on the basis of isotropic and anisotropic elasticity theories using elastic constants measured at room temperature.

5.2.2 Film-Single Crystal Epitaxy

Under ultrahigh-vacuum conditions epitaxial single crystal CoSi$_2$ can be obtained from deposited Co on Si (111). This CoSi$_2$ film is termed B, contrary to the so-called A film which is obtained in a non-ultrahigh-vacuum condition. Examples of these two kind dark field CoSi$_2$ images can be seen in Fig. 5.9. The deposition of Co was done by electron beam deposition (EB). Both Fig. 5.9a, b were taken in a direction

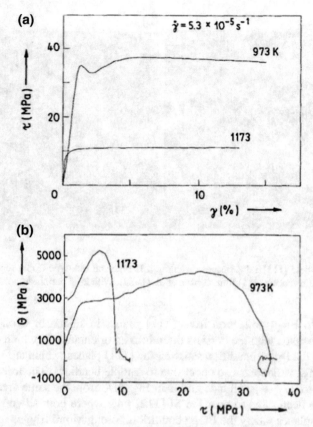

Fig. 5.8 Compression test results at 973 and 1173 K in regime 2 for $\dot{\gamma} = 5.3 \times 10^{-5}$ s^{-1}. **a** Stress-strain curves and **b** the associated rates of hardening. Anongba and Steinemann (1993). With kind permission of John Wiley and Sons

Table 5.1 Stacking fault energy $\bar{\gamma}$ and partial separation associated with dissociated dislocations of various slip systems. Anongba and Steinemann (1993). With kind permission of John Wiley and Sons

Slip system	Dislocation character	Δ_{obs} (10^{-10} m)	Δ (10^{-10} m)	$\bar{\gamma}$ (mJ m^{-2})	
				Anisotropic elasticity	Isotropic elasticity
[010](001)	62°	59 ± 5 g = [020]	53 ± 6	181 ± 21	175 ± 38
	90°	58 ± 5 g = [020]	51 ± 6	194 ± 24	194 ± 43
1/2[0$\bar{1}$1](011)	28°	37 ± 7 g = [0$\bar{2}$2]	38 ± 8	36 ± 7	34 ± 11
1/2[110]($1\bar{1}$1)	13°	35 ± 7 g = [020]	38 ± 8	39 ± 6	45 ± 13

(a)   **(b)**

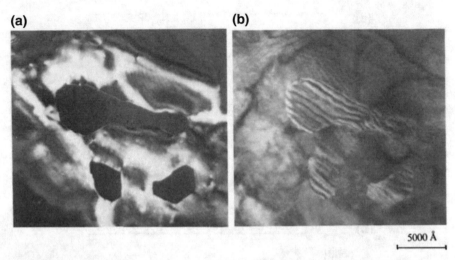

5000 Å

Fig. 5.9 Dark field $(11\bar{1})$ TEM images revealing **a** B-type and **b** A-type CoSi$_2$ regions in a non-ultrahigh-vacuum-reacted (111) film. Gibson et al. (1982). With kind permission of Elsevier

appropriate to Bragg reflection from $(11\bar{1})$ planes in silicon or CoSi$_2$. However, Fig. 5.9a was taken with the crystals tilted into an orientation in which only B-type regions have the Bragg condition satisfied for $(11\bar{1})$ planes, which are bright in this image (contrast variations also occur due to sample bending). The three dark areas are revealed to be A-type silicide grains in Fig. 5.9b from the same area, where the crystals have been tilted to near the Si [112] pole where both silicon and A-type CoSi$_2$ $(11\bar{1})$ planes satisfy the Bragg condition. Strong Moiré fringes in the silicide are seen which comes from contrast in other areas from the silicon substrate. By tilting thin samples into weak beam conditions, the interface dislocation networks can be imaged in the electron microscope. In Fig. 5.10 these dislocations are seen at the interface of a 600 Å thick ultrahigh-vacuum-codeposited film. Three types of defects are seen: (1) A regular hexagonal network of edge dislocations with Burgers vector of type $1/6[11\bar{2}]$ in $[1\bar{1}0]$-type directions (only two of three segments are visible in any $(2\bar{2}0)$-type reflection); (2) a triangular network of dislocations in $[1\bar{1}0]$-type directions irregularly spaced; (3) occasional (less than 10^8 cm^{-2}) dislocations inclined to the interface plane with Burgers' vectors of type $1/2[110]$. The type (3) act as sources for pairs of type (1) dislocations and their glide can relief of small amounts of misfit stress, which probably occurs during cooling from the reaction or deposition temperature.

A-type CoSi$_2$ grains have an entirely different defect structure, namely more irregular and widely spaced than type 1 dislocations. It comprises edge dislocations with Burgers' vectors of type $1/2[1\bar{1}0]$ in $[11\bar{2}]$-type directions. Dislocations of the second kind are not found under type A islands. The only microstructural distinction in the type B grains is the type 2 defects. They are extremely straight and very long, often running for more than 10 μm. They are consequently parallel to the CoSi2–Si interface plane and tilting experiments indicate them to be within 50 Å or less of

Fig. 5.10   Weak beam $(\bar{2}20)$ TEM image from a 600 Å thick ultrahigh-vacuum-codeposited CoSi₂ film. Revealed is the characteristic defect structure found everywhere beneath this B-type single crystal. Gibson et al. (1982). With kind permission of Elsevier

this plane. Because of their strong interaction with type I interface misfit dislocations they seem to be at the interface itself. The interfacial dislocation structure is shown in a plan view in Fig. 5.11, which is a weak beam dark field TEM from MBE CoSi₂ film codeposited at 650 °C. A regular network of interfacial misfit dislocations spaced

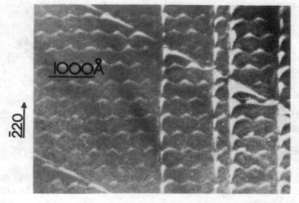

Fig. 5.11   A $(2\bar{2}0)$ dark field image of a 650 Å codeposited CoSi₂ layer seen in plan view. Tung et al. (1982). With kind permission of Elsevier

by about 350 Å apart is seen. Only two sets of hexagonal network is revealed by this $(2\bar{2}0)$ Bragg image. The Burgers vectors of these dislocations are of $1/6(1\bar{2}1)$ type and are pure edge dislocations in the interface plane and are allowed only in the B type interface, since they are not perfect dislocations in FCC lattice. The intense extremely straight lines are always seen at type B CoSi$_2$ interfaces, but not found at type A CoSi$_2$.

5.3 Dislocations in NiSi$_2$

5.3.1 Epitaxial Thin Film

Epitaxial thin films were formed on silicon substrate. The interface planes were faceted as revealed by TEM bright field and dark field imaging. Epitaxial growth was found on all low index planes{111}, {011} and {001}. The mismatch of 0.5% between NiSi$_2$ and silicon indicates closely matching lattice parameters between them. The transformation of the polycrystalline silicide to the epitaxial one was studied by in situ annealing (850 °C) of nickel thin film (300–400 Å) on silicon in the electron microscope. The nickel was evaporated at room temperature by electron gun onto (001) and (111) oriented silicon in a vacuum better than 1×10^{-7} torr. Figure 5.12 illustrate bright field and dark field micrographs of specimens with Si (100) orientation.

The epitaxial silicide is known to be faceted with {111} and {001} interfaces. Ledges several atomic planes high are revealed by the dark field images and show that the interfaces between NiSi$_2$ and silicon are mainly {111} which have lower interface energy than {001} interfaces.

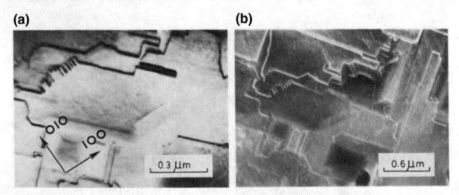

Fig. 5.12 a Bright field and **b** dark field micrographs showing the structure of epitaxial NiSi$_2$ on Si (001) through the strain contrast due to lattice mismatch at the interface. Chen et al. (1982). With kind permission of Elsevier

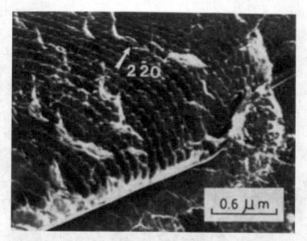

Fig. 5.13 Micrograph showing the configuration of interface dislocation networks of the NiSi₂/Si⟨111⟩ system for a sample annealed at 800 °C for 1 h. Chen et al. (1982). With kind permission of Elsevier

At an annealing of 800 °C for 1 h epitaxial film was formed but TEM has indicated that the film is mainly polycrystalline. The polycrystalline film transforms later (820 °C) to the epitaxial NiSi₂. Square, hexagonal and irregular dislocation networks were observed. The dislocation configuration is seen in Fig. 5.13. Analysis indicate that the dislocations are edge type with $1/6\langle 112\rangle$ Burgers vectors. Edge dislocations with $1/2\langle 110\rangle$ vectors were also observed. The average spacing of the dislocations is ~1000 Å, sufficiently close to the theoretically expected value of 970 Å. Dislocation networks at the silicide-silicon interface were analyzed by the diffraction contrast method.

In a relatively more recent publication epitaxial NiSi₂ formation was explored in-situ by reaction of single crystalline silicon and nickel. The in-situ investigation showed the sequence of phase formation—which is temperature dependent—during the reaction of nickel particles with single crystalline Si (100) on annealing. At annealing above 600 °C formation of NiSi₂ occurs accompanied by its epitaxial growth. The growth of crystalline NiSi₂ is accompanied with formation of dislocations both in the NiSi₂ and silicon phases. The temperature of the silicide formation epitaxially in a microscope can be reduced to 400 °C if the reaction is performed with irradiation by a beam of accelerated electrons.

Annealing at 700 °C of Ni particles on silicon results only in the formation of NiSi₂ and its growth. The increase in size of the growing NiSi₂ phase is accompanied by the appearance of contrast characteristic of dislocations as illustrated in Fig. 5.14. A decrease in temperature even of 50 °C causes some decrease in the number of dislocation lines in the NiSi₂ phase. Dislocation formation with the growth of the NiSi₂ phase does not occur only in it, but also on the NiSi₂–Si interface in the silicon phase also. This can be seen in Fig. 5.15. The movement of dislocations in

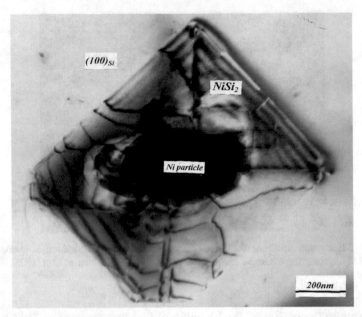

Fig. 5.14 TEM microphotograph of the dislocations in NiSi$_2$ crystals formed during annealing of the system: nickel particle/single crystalline silicon film at 700 °C. Bokhonov and Korchagin (2001). With kind permission of Elsevier

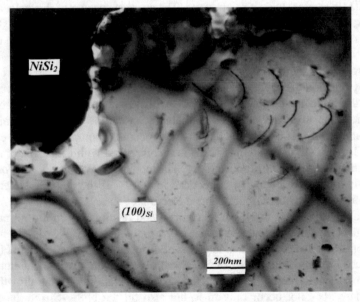

Fig. 5.15 TEM microphotograph of dislocations formed in the silicon phase near the NiSi$_2$–Si interface. Bokhonov and Korchagin (2001). With kind permission of Elsevier

the disilicide and silicon phase during formation of the NiSi$_2$ phase was observed directly during in-situ electron microscopic investigation.

5.4 Dislocations in MoSi$_2$

5.4.1 Single Crystals

In the following section the dislocation structure observed in MoSi$_2$ single crystals following deformation along the soft $\langle 110 \rangle$ compression axis in the temperature range of room temperature-1050 °C is considered. The specimens for the compression tests were cut from MoSi$_2$ single crystals with a [110] compression axis and (110) and (001) side faces. The strain rate sensitivity of the flow stress was determined by stress relaxation tests. The stress strain curves with several stress relaxations R$_1$–R$_5$ and a temperature change is seen in Fig. 5.16. The temperature dependence of the flow stress shows three temperature ranges, a low and a high-temperature range with a normal decrease of the flow stress with increasing temperature and an intermediate range with an anomalous flow stress increase. High voltage electron microscope (HVEM) was used to look at the dislocation structures of the deformed samples which were imaged in diffraction contrast. The specimens for transmission electron microscopy were cut from the deformed samples parallel to the (110) cross section plane or to the (001) or (1$\bar{1}$0) side faces. The dislocation structure was investigated in the (HVEM) at an acceleration voltage of 1000 kV using high-order bright field diffraction contrast. In a number of cases, the Burgers vectors of the dislocations were determined by the $\mathbf{g} \cdot \mathbf{b} = 0$ contrast extinction rule. An example is presented in Fig. 5.17 with a dislocation structure near a precipitate. The dislocations were generated by deformation at 380 °C.

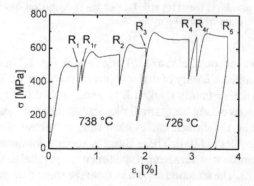

Fig. 5.16 Stress-strain curve of a deformation experiment at 738 and 726 °C. Strain rate 10^{-5} s^{-1}. R$_n$ stress relaxation tests and R$_{nr}$ repeated relaxation tests. Dietzsch et al. (2005). With kind permission of John Wiley and Sons

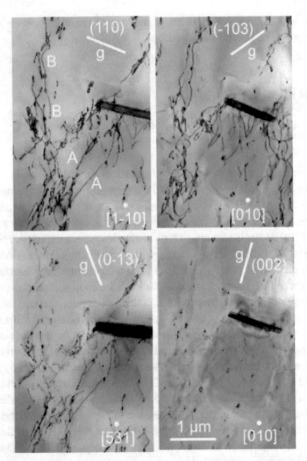

Fig. 5.17 Microstructure after deformation along ⟨110⟩ at 380 °C by 0.4% to final stress of 400 MPa imaged with different **g** vectors at different poles. The poles are indicated by the indexing at the white dot. Foil plane parallel to the (110) side face of the compression specimen. Dietzsch et al. (2005). With kind permission of John Wiley and Sons

The specimen was cut parallel to a $(1\bar{1}0)$ side face. In Fig. 5.17a with $\mathbf{g} = (110)$, all dislocations are visible. A family of dislocations marked A is invisible in Fig. 5.17b with $\mathbf{g} = (103)$. Another family marked B is extinguished in Fig. 5.17c with $\mathbf{g} = (0\bar{1}3)$. All dislocations except some small dislocation loops are extinct in Fig. 5.17d with $\mathbf{g} = (002)$. Thus, the dislocations A have Burgers vectors parallel to [010] and those labelled B parallel to [100]. These Burgers vectors dominate in all specimens studied. The dislocation structure from a specimen cut parallel to the (001) side face is shown in Fig. 5.18. The structure shown is from the same deformation experiment as that of Fig. 5.17.

In this projection, all four {101} slip planes are inclined to the foil plane unlike in Fig. 5.17b, d where two of the four slip planes were imaged. It is observed that

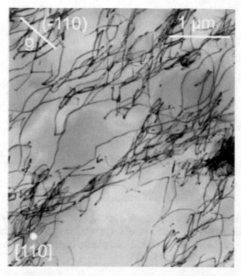

Fig. 5.18 Dislocation structure of the same deformation experiment as in Fig. 15.17. Foil plane parallel to (110) cross section plane. Dietzsch et al. (2005). With kind permission of John Wiley and Sons

the dislocations are curved at this temperature (end of the normal low temperature range of the flow stress). The density of the dislocation was evaluated by

$$\rho = 2\,\text{N/Lt} \tag{5.1}$$

where L is the total length, and t is the thickness of the foil. The conditions of the deformation of the specimen shown in Fig. 5.18 are: temperature 380 °C, plastic strain 0.4%. The value of ρ is 2.8×10^{13} m^{-2}.

The dislocation structure of specimens deformed at 606 °C are illustrated in Figs. 5.19 and 5.20. The foils are cut parallel to (001) side face of the compression. In Fig. 5.19 the imaging was done near the $[\bar{1}\bar{1}1]$ pole with $\mathbf{g} = (1\bar{1}0)$, thus all dislocations with ⟨110⟩ Burgers vectors are visible. Many of the dislocations are straight and orient parallel to the traces of (101) and (011) planes. The microstructure of Fig. 5.20 is the result of imaging near the [001] pole with a $\mathbf{g}$ vector which in Fig. 5.20a is of $\mathbf{g}(1\bar{1}0)$, thus all the dislocations are visible. In Fig. 5.20b a set of dislocations are extinguished with $\mathbf{g} = (020)$. Further another set of dislocations are extinguished in Fig. 5.20c with $\mathbf{g} = (200)$. The Burger vectors are also ⟨100⟩. Most of the dislocations in this figure are also straight and are oriented parallel to $[1\bar{1}0]$ or perpendicular to the [110] direction. Thus, these dislocation lines are oriented along ⟨111⟩ directions. The residual contrast at the extinctions with $\mathbf{g} \cdot \mathbf{b} = 0$ (Fig. 5.20b, c) is relatively large. Dislocations with other orientations are also imaged along the traces of these planes. Dislocations along the remaining ⟨111⟩ direction on the {101}

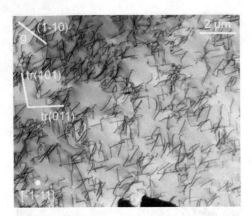

Fig. 5.19 Dislocation structure after deformation along ⟨110⟩ at 606 °C by 0.2% to final stress of 350 MPa. Foil plane parallel to (001) side face of compression specimen. The straight lines marked tr indicate the directions of the traces of the indicated planes imaged edge-on. Dietzsch et al. (2005). With kind permission of John Wiley and Sons

planes which are not imaged edge-on are visible in diagonal direction between the former ones.

Specimens deformed at different temperatures and at higher stresses and strain are illustrated in Fig. 5.21. In Fig. 5.21a the dislocations are near the [111] pole while in Fig. 5.21b similar area is shown near the [001] pole. The dislocation structure is less regular than that at the smaller strain. They form a cell structure with bundles which appear also parallel to [110] direction. The bundles may form by the interaction between dislocations on the two {101} planes which intersect along the [111] direction.

Figures 5.22 and 5.23 illustrate dislocation structures formed at higher temperature of 672 °C (within the earlier indicated anomaly range). Figure 5.22 is a stereo pair taken from the $[1\bar{1}0]$ pole. The compression axis is parallel to the (110) **g** vector. All the four {101} slip planes are inclined to the foil plane at the same equal angles. It is seen in the images that the dislocations have angular shapes (with segments in the projection of the ⟨111⟩ direction). In Fig. 5.23 the same structure at the $[\bar{3}\bar{3}1]$ and [100] poles is seen. In Fig. 5.23a, b, the two {301} reflections extinguish one set of dislocations each. In Fig. 5.23c taken with the (002) reflection, all dislocations are extinguished.

The Burgers vectors are again parallel to [100] and [010] (they appear only in a relatively strong residual contrast). In this projection, practically all dislocations are parallel to the projections of the four ⟨111⟩ directions. At high temperature deformation (738/726 °C) the dislocation structure is less regular as can be seen in Fig. 5.24. Only (011) planes of the four {101} slip planes are oriented edge on in Fig. 5.24. The dislocation bow smoothly out between cusps. There are many small loops probably formed from the dislocation debris at the high temperature. In several cusps loops are situated and act as obstacles. The microstructure is the result of deformation along ⟨110⟩. Higher dislocation density in a band structure can be

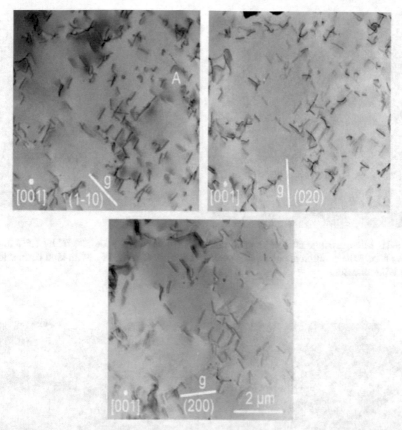

Fig. 5.20 Microstructure of the same specimen as Fig. 5.19 at different **g** vectors near the [001] pole. Dietzsch et al. (2005). With kind permission of John Wiley and Sons

seen in Fig. 5.25. Different slip systems are operating. The bands are indicated by the letters A, B and C. The borders of the band are in [010] direction. In the band A the projection of many dislocations run in the two $\langle 110 \rangle$ directions and in $[\bar{1}10]$ direction in band B. In Fig. 5.25a, b dislocations are visible both with the (110) and (020) **g** vectors. They are however extinguished with **g** = (200) in Fig. 5.25c. Their Burgers vector should be parallel to [010]. The increased dislocation density in the borders may have formed by the interaction between dislocation on (101) and $(\bar{1}01)$ planes.

The results of in situ experiments are illustrated in Fig. 5.26. Here the samples were predeformed at 495, 606 and at 907 °C. The in situ test were carried out at 475, 610, 835, and 970 °C. The in situ experiments confirm the results seen earlier in the ex-situ experiments.

In summary of the plastic deformation of MoSi$_2$ single crystals along $\langle 110 \rangle$ one can arrive to the conclusion that plastic deformation is carried out by dislocations with $\langle 100 \rangle$ Burgers vectors moving on $\{011\}$ planes. Regarding the temperature

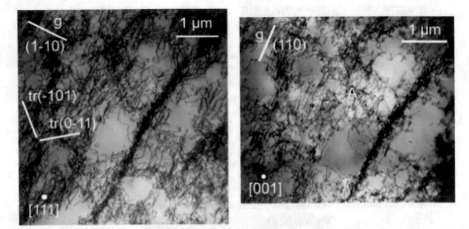

Fig. 5.21 Microstructure after deformation along ⟨110⟩ at 638/622/606/590 °C by 1.5% to final stress of 750 MPa at different poles and **g** vectors. Dietzsch et al. (2005). With kind permission of John Wiley and Sons

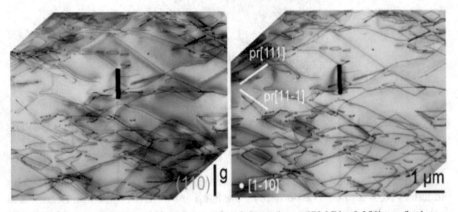

Fig. 5.22 Stereo pair of dislocation structure after deformation at 672 °C by 0.25% to a final stress of 450 MPa. The straight lines marked pr indicate the directions of the projections of the indicated orientations. Dietzsch et al. (2005). With kind permission of John Wiley and Sons

dependence of the flow stress three ranges can be outlined for the {110}1/2⟨111⟩ slip system: (1) A low-temperature range with decreasing flow stress where the behavior can be interpreted by the action of the Peirels mechanism superimposed with a large athermal flow stress component. (2) An anomaly range with an increasing flow stress, where the idea is that the dislocations dissociate along certain crystallographic directions by glide and conservative climb and (3) a high temperature range with normally decreasing flow stress where the flow stress is controlled by recovery.

The functions applied at the different ranges can be indicated as:

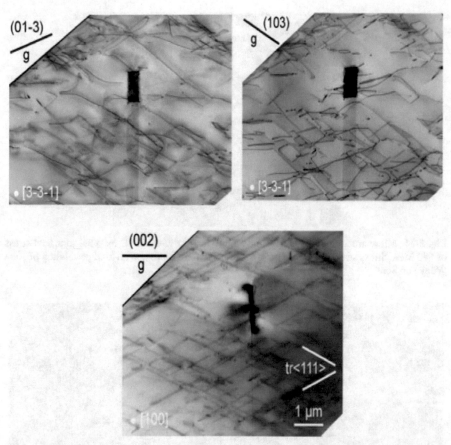

Fig. 5.23 Same dislocation structure as in Fig. 5.22 at different poles and **g** vectors. Dietzsch et al. (2005). With kind permission of John Wiley and Sons

(a) Low temperature range: The Peierls mechanism—dependence of the Gibbs free energy of activation on the effective stress τ^* for the double kink mechanism

$$\Delta G(\tau^*) = 2G_k \left[1 - \frac{\pi\tau^*}{8\tau_p} \left(\ln \frac{16\tau_p}{\pi\tau^*} + 1 \right) \right]$$

(5.2)

G is the energy of single kink, τ_p the Peierls stress at $T = 0$ K. The activation volume from this potential is

$$V(\tau^*) = -\frac{\partial \Delta G}{\partial \tau^*} = \frac{\pi G_k}{4\tau_p} \ln \frac{16\tau_\pi}{\pi\tau^*}$$

(5.3)

Fig. 5.24 Microstructure after deformation along ⟨110⟩ at 738/726 °C by 3.5% to a final stress of 680 MPa. Stress strain curve in Fig. 5.16. Dietzsch et al. (2005). With kind permission of John Wiley and Sons

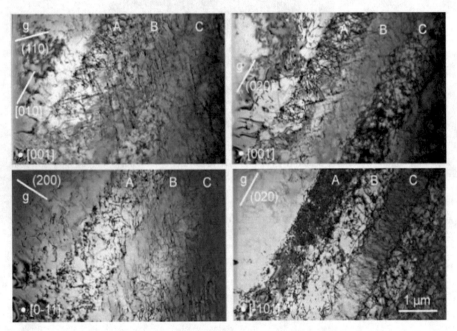

Fig. 5.25 Microstructure of the same specimen as in Fig. 5.23 at different poles and **g** vectors. Dietzsch et al. (2005). With kind permission of John Wiley and Sons

an analysis of V versus τ^* is not possible. Therefore, Eqs. (5.2) and (5.3) together with the Arrhenius equation

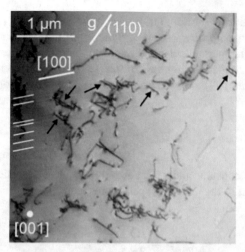

Fig. 5.26 Dislocations during in situ straining in an HVEM at 475 °C along $\langle 110 \rangle$ of a specimen predeformed at 606 °C. Dietzsch et al. (2005). With kind permission of John Wiley and Sons

$$\dot{\varepsilon} = \dot{\varepsilon}_0 \exp\left(-\Delta G / kT\right) \tag{5.4}$$

are solved for T. $\dot{\varepsilon}$ is the strain rate and $\dot{\varepsilon}_0$ is a preexponential factor. Assuming $\dot{\varepsilon}_0$ is constant T is given as

$$T = T_0[1 - 2\exp(-\chi V)(\chi V + 1) \tag{5.5}$$

where

$$T_0 \equiv \frac{2G_k}{k \ln(\dot{\varepsilon}_0/\dot{\varepsilon})} \text{ and } \chi \equiv \frac{4\tau_p}{\pi G_k} \tag{5.6}$$

Regression analysis of T versus V yields the values for $T_0 = 554$ K and $\chi = 4.71$ nm^{-3}. With $\ln(\dot{\varepsilon}_0/\dot{\varepsilon}) = 20$, it follows that $G = 0.48$ eV and $\tau_p - 282$ MPa.

(b) The anomaly range: In MoSi₂ dissociation is possible of the dislocations with $1/2\langle 111 \rangle$ Burgers vector along $\langle 110 \rangle$ as

$$1/2[111](1\bar{1}0) \rightarrow 1/6[331](1\bar{1}0) + 1/6[331](\bar{1}16) + 1/6[\bar{3}31](1\bar{1}6) \tag{5.7}$$

This dissociation is favored by a small gain in elastic energy and because $1/\langle 111 \rangle$ Burgers vectors seem generally not to be vary stable. The total dislocation can move on its glide plane only by a combination of glide and conservative climb (i.e. emission of vacancies by one partial dislocation and the absorption of these vacancies by another partial dislocation. The dynamic behavior is considered by the kinetics of the Cottrell (1953) effect which can be used as a first approximation. The dislocation does not constrict during motion, but moves by a combination of glide and climb, which require diffusion processes. Since a theory of the conservative climb model with velocity and temperature depending dissociation width is not yet available, some estimations are made using the Cottrell (1953) effect model mentioned. The temperature maximum T_{max} for the flow stress maximum is given by

$$T_{max} = \frac{\dot{\varepsilon}}{4\rho bkD}\beta \tag{5.8}$$

β is an interaction constant and as usual ρ and D the dislocation density and the diffusion coefficient of the diffusing species, respectively.

$$\beta = \frac{\mu b}{3\pi}\frac{1+\nu}{1-\nu}\nu_{rel} \tag{5.9}$$

μ is the shear modulus, which for the Cottrell (1953) effect is replaced by the anisotropic K value, K_e for the edge dislocation. ν is the Poisson's ratio and for 650 °C $K_e = 187.2$ GPa. $\nu = 1 - E_s/E_e$ where the E's are the prelogarithmic energy factors of the screw and edge dislocations, respectively; ν is very small: $\nu = 0.04$. In Eq. (5.9) ν_{rel}, is the relaxed volume of a vacancy. It is assumed here that ν_{rel} is ~30% of the atomic volume namely $\nu_{rel} = 0.3$ a^{2}c/6 with a and c the lattice constants. The resulting β with these values is $\beta = 2.8 \times 10^{-29}$ Nm^2. Then, it follows from Eq. (5.8) that the diffusion coefficient should be about $D \cong 5 \times 10^{-19}$ m^2 s^{-1} to obtain the flow stress maximum at a strain rate $\dot{\varepsilon} = 10^{-5}$ s^{-1} and a temperature of $T_{max} = 1010$ K. The maximum value of the Cottrell effect contribution is estimated as

$$\Delta\sigma_{c,max} = \frac{17c\beta}{m_s b^4} \tag{5.10}$$

where c is the concentration of the diffusing defect and $\Delta\sigma_{c,max} = 280$ MPa. The necessary vacancy concentration is $c = 3 \times 10^{-3}$.

In addition to the mechanism causing the flow stress anomaly, there is also a large athermal flow stress component due to long-range dislocation interactions. It dominates at the flow stress minimum between 300 and 500 °C.

(c) The high temperature range: Above the flow stress peak the flow stress decreases normally. Band structure was observed (Fig. 5.25). Long range dislocations interactions occur and the flow stress is controlled by recovery. Since work hardening is low the equation of steady state creep can be applied as

$$\bar{\dot{\varepsilon}} = A\frac{\mu b}{kT}\left(\frac{\sigma}{\mu}\right)^m D_0\left(\frac{\Delta H}{kT}\right) \tag{5.11}$$

A is a dimensionless constant m is the stress exponent (used in creep) D_0 pre-exponentials factor and ΔH the activation enthalpy. Depending on the creep model $m = 4 - 6$. The stress exponent can be determined from the strain rate sensitivity, r according to

$$m = \frac{d \ln \dot{\varepsilon}}{d \ln \sigma} = \frac{\sigma}{r} \tag{5.12}$$

5.5   Dislocations in WSi₂

5.5.1   Single Crystals

Compressive stress strain curves, yield stress as a function of temperature and the CRSS were illustrated in Figs. 4.33, 4.35 and 4.36 for WSi₂ of various orientations. The dislocations associated with the deformations are illustrated below: In Fig. 5.27 typical dislocation structures are seen for slip on (001) in crystal oriented in $[\bar{1}12]$ deformed at 1300 °C. The thin foil was cut parallel to the (001) slip plane. Long dislocations are seen which tend to align along [100] and [010] directions. In Fig. 5.28 contrast analysis is performed to determine their Burgers vector. The dislocations aligned along [010] direction taken with the reflection vector shown in Fig. 5.28a, Fig. 5.28b and Fig. 5.28d are invisible for $\mathbf{g} = 110, \mathbf{g} = 03\bar{3}$ and $\mathbf{g} = 020$, respectively, but they are in contrast for other reflection in Fig. 5.28 yielding the Burgers vector $\mathbf{b} = [100]$. Dislocations aligned along the [100] in Fig. 5.28a, Fig. 5.28c and Fig. 5.28e

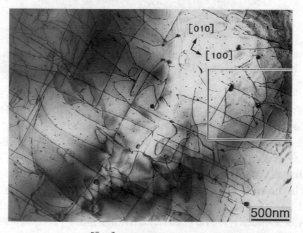

Fig. 5.27 Dislocation structure in a $[\bar{1}12]$-oriented single crystal deformed at 1300 °C. The thin foil was cut parallel to the (001) macroscopic slip planes. The crystallographic directions are indicated in the figure. Ito et al. (1999). With kind permission of Elsevier

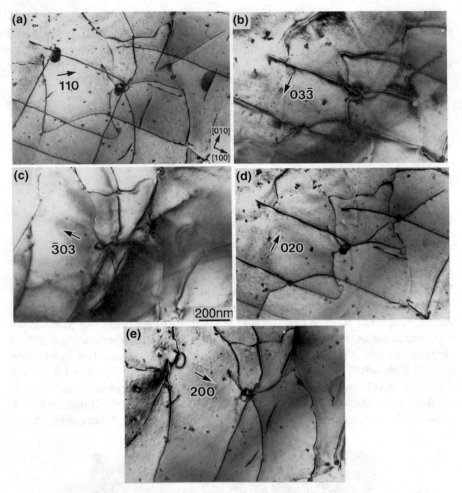

Fig. 5.28 Contrast analysis of dislocations in the outlined area in Fig. 5.27. The operating reflections are indicated in the figures. Ito et al. (1999). With kind permission of Elsevier

are invisible for $\mathbf{g} = \bar{3}03$ and $\mathbf{g} = 200$ respectively, yielding the Burgers vector $\mathbf{b} = [010]$. No appreciable dissociation of (100) dislocations gliding on (001) has occurred. Thus the slip direction for slip on (001) is (100). Dislocations with $\mathbf{b} = (100)$ tend to align along their edge orientation. (For slip consult Fig. 4.37).

Contrast analysis made to determine the Burgers vector of dislocations gliding on {110} planes is shown in Fig. 5.29. Crystals were cut from [0 15 1] oriented crystal deformed at 1200 °C to about 15 strain. The thin foil was cut parallel to (010) slip planes. Long dislocations lying on (110) tend to align along $[\bar{1}10]$ as seen in Fig. 5.29a taken with $\mathbf{g} = \bar{1}10$. These dislocations are invisible for $\mathbf{g} = \bar{2}1\bar{3}$ and $\mathbf{g} = 1\bar{2}3$ of Fig. 5.29f, g but visible for other reflections in Fig. 5.29, yielding the Burgers vector $\mathbf{b} = 1/2[\bar{1}11]$. Thus the slip system identified is {110}⟨111⟩. The short dislocation

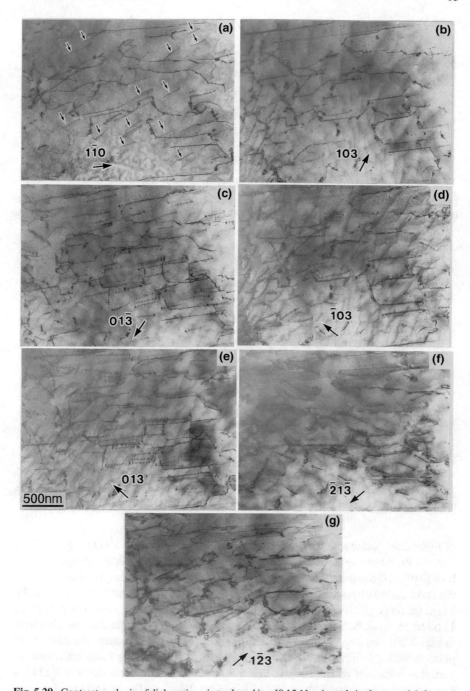

Fig. 5.29 Contrast analysis of dislocations introduced in a [0 15 1]-oriented single crystal deformed at 1200 °C. The thin foil was cut parallel to (110) macroscopic slip planes. The operating reflections are indicated in the figures. Ito et al. (1999). With kind permission of Elsevier

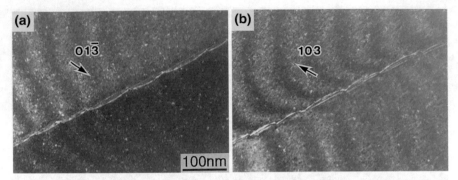

Fig. 5.30 Weak-beam images of a $1/2[\bar{1}11]$ dislocation observed in the foil of Fig. 5.29. The operating reflections are indicated in the figures. Ito et al. (1999). With kind permission of Elsevier

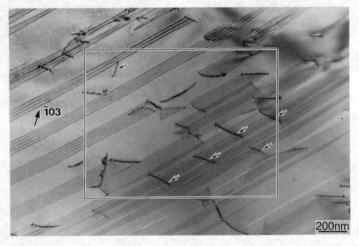

Fig. 5.31 Deformation structure in a [001]-oriented single crystal deformed at 1500 °C. The thin foil was cut parallel to (013) planes. Ito et al. (1999). With kind permission of Elsevier

of Fig. 5.29a (some marked with arrows) are invisible for $\mathbf{g} = 103$ and $\mathbf{g} = \bar{1}03$ in Fig. 5.29b, d but visible for other reflections of Fig. 5.29 yielding a Burgers vector $\mathbf{b} = [010]$. Tilting the three zone axis orientations and trace analysis indicated that the short dislocations are aligned parallel to [100] suggesting that they are on (001). Thus, the (001)⟨100⟩ slip system also contributes to the plastic deformation of [0 15 1] oriented crystals. Stacking faults were shown in Fig. 4.34, but they are also seen in Fig. 5.29f and Fig. 5.29g ($\mathbf{g} = \bar{2}13$ and $1\bar{2}3$, respectively) where the (001) fault plane is inclined with respect to the incident beam. The stacking faults are bound by partial dislocations of 1/3[001]. In Fig. 5.30 the dissociation scheme of $1/2[\bar{1}11]$ is imaged of the specimens illustrated in Fig. 5.29. Dissociation of $1/2[\bar{1}11]$ into two partials is observed when imaged with $\mathbf{g} = 01\bar{3}$ and $\mathbf{g} = 103$ with zone axes [331] (Fig. 5.30a) and with $[33\bar{1}]$ (Fig. 5.30b), respectively. Thus 1/2[111] dislocations dissociate into collinear partials separated by a superlattice intrinsic stacking fault (SISF) according to

$$1/2[111] \rightarrow 1/4[111] + SISF + 1/4[111] \qquad (5.13)$$

Deformation structure in [001] oriented crystals deformed at 1500 °C to about 1% strain is seen in Fig. 5.31. There are much more stacking faults in the deformed crystal than in the as grown one shown in Fig. 4.34. The arrows in the figure indicate dislocations not bound by stacking faults. Contrast analysis results to determine the Burgers vectors of the dislocations in outlined area of Fig. 5.31 is shown in Fig. 5.32.

Dislocation segments not bound by stacking faults shown in Fig. 5.32a (marked by arrows) are invisible for $\mathbf{g} = 01\bar{3}$ and $\mathbf{g} = 020$ (see Fig. 5.32e and Fig. 5.32f) respectively, but are visible for other reflections, yielding the Burgers vector $\mathbf{b} = [100]$. Their slip plane by trace analysis indicate as being (001).

Stacking faults on (011) are visible for $\mathbf{g} = \bar{1}03$, $\mathbf{g} = 103$ and $\mathbf{g} = 01\bar{3}$ in Fig. 5.32a, Fig. 5.32b and Fig. 5.32c, respectively, and are invisible for other reflections. This indicates that translation vector is parallel to [001]. The grown in stacking faults of Fig. 4.34 are bounded by partial dislocations with Burgers vector $\mathbf{b} = 1/3[001]$; the stacking faults imaged in Fig. 5.32 are different. Bounding dislocations A$_1$ and A$_2$ seen in Fig. 5.32a have Burgers vectors $\mathbf{b} = 1/6[3\bar{3}1]$ and $1/6[33\bar{1}]$ since they are invisible for $\mathbf{g} = \bar{1}03$, $\mathbf{g} = 110$, $\mathbf{g} = 103$ and $\mathbf{g} = 1\bar{1}0$ shown in Fig. 5.32a, Fig. 5.32d, Fig. 5.32b and Fig. 5.32c, respectively. Dislocations labeled A$_3$ and A$_4$ have Burgers vectors $\mathbf{b} = 1/6[331]$ and $1/6[3\bar{3}\bar{1}]$, respectively are considered to result from the dissociation of $\langle 100 \rangle$ dislocations by the equation of

$$[100] \rightarrow 1/6[3\bar{3}1] + APB + 1/6[33\bar{1}] \qquad (5.14)$$

APB stands for antiphase boundary. In the above reaction climb process is involved of the partial or the partial dislocations. The (001) stacking faults bounded by the above partials are formed during deformation through a climb process contrary to those formed during crystal growth.

The operating slip systems in WSi$_2$ were indicated in Table 4.7 as (001)$\langle 100 \rangle$, {011}$\langle 100 \rangle$, {023}$\langle 100 \rangle$ and{110}$\langle 111 \rangle$ in the temperature range from 1100 to 1500 °C. WSi$_2$ can be deformed only above 1100 °C. For a summary on deformation of WSi$_2$ the reader is referred to Sect. 4.3.3.

5.6   Dislocations in TiSi$_2$

5.6.1   Single Crystals

Although compressive deformation in single crystal C54 TiSi$_2$ has been illustrated earlier in Figs. 4.38, 4.39, 4.40, 4.41 and 4.42, below compression and slip will be

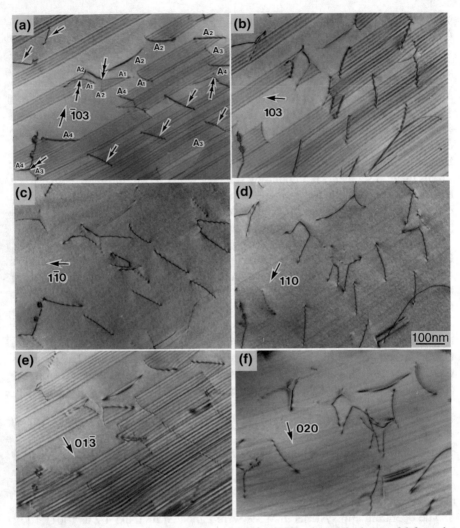

Fig. 5.32 Contrast analysis of dislocations introduced in a [001]-oriented single crystal deformed at 1500 °C. The operating reflections are indicated in the figures. Ito et al. (1999). With kind permission of Elsevier

reconsidered since the dislocation structure is related to the deformation information presented below. The deformation illustrations in compressions are also presented as a function of crystal orientation in the temperature range of RT −1400 °C. The orientations of the compression axes, being [001], [100], [010], [110], [101] and [021] are shown in Fig. 5.33.

The yield stress of specimens with orientation of [101] and [021] are shown in Fig. 5.34 as a function of temperature.

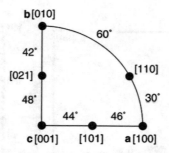

Fig. 5.33 Stereographic projection of compression axis orientations investigated. Inui et al. (2003). With kind permission of Elsevier

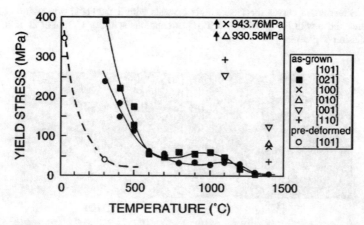

Fig. 5.34 Temperature dependence of yield stresses for TiSi₂ single crystals with [101] and [021] orientations. Data points plotted with symbols ∇, ×, Δ and + indicate the stresses at which specimen buckling occurred for [001], [100], [010], [110] orientations, respectively. Inui et al. (2003). With kind permission of Elsevier

The buckling of specimens occurred at the stresses indicated in Fig. 5.34 with the symbols ∇, ×, Δ and +. Following the decrease in yield stress with temperature a plateau-like region is observed above ~700 °C. Stress-strain curves for [101] orientation are shown in Fig. 5.35. A yield drop can be seen at the lower temperatures, the magnitude of which decreases with increasing temperatures. Serrated stress-strain curves are seen. Similar curves were observed in [021] orientation. Slip lines of TiSi₂ deformed at 500 °C are illustrated in Fig. 5.36 for the orientations indicated.

Table 5.2 lists the Schmidt factors for the orientations [001], [100], [010], [110], [101] and [021] mentioned above. The critical resolved shear stress (CRSS) for slip on (001)⟨110⟩ is calculated as a function of temperature and the results are seen in Fig. 5.37.

Dislocations of TiSi₂ deformed at 500 °C belonging to (001)⟨110⟩ are seen in Fig. 5.38. The dislocations of Fig. 5.38c are invisible of $\mathbf{g} = \bar{3}3\bar{1}$ (Fig. 5.39b) and

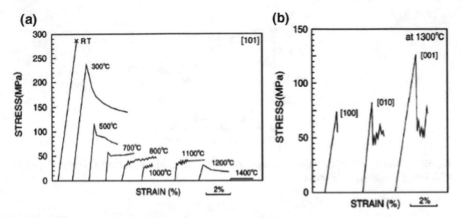

Fig. 5.35 Stress-strain curves for TiSi$_2$ single crystals with **a** the [101] orientation at selected temperatures and **b** with [100], [010] and [001] orientations at 1300 °C. Inui et al. (2003). With kind permission of Elsevier

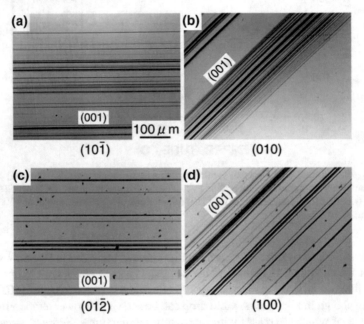

Fig. 5.36 Slip lines observed on two orthogonal faces of **a**, **b** [101]- and **c**, **d** [021]-oriented TiSi$_2$ single crystals deformed at 500 °C. The compression axis is in the direction parallel to the vertical edge of the figures. Inui et al. (2003). With kind permission of Elsevier

$\mathbf{g} = \bar{3}31$ (Fig. 5.38d). Dislocations seen in Fig. 5.38b are invisible for $\mathbf{g} = \bar{3}\bar{3}1$ (Fig. 5.38c) and $\mathbf{g} = \bar{3}3\bar{1}$ (Fig. 5.38e) yielding Burgers vectors $\mathbf{b} = 1/2[110]$ and $1/2[110]$ respectively. Recall that $\mathbf{g}$ is the reflection vector.

Table 5.2 The largest Schmid factors for the slip systems previously reported on TiSi$_2$ single crystals with six different orientations investigated. Inui et al. (2003). With kind permission of Elsevier

Slip system	Orientation [100]	[010]	[001]	[110]	[101]	[021]
(001)⟨110⟩	0	0	0	0	0.432	0.249
(001)[010]	0	0	0	0	0	0.497
{01̄1̄}⟨011⟩	0	0.426	0.426	0.107	0.220	0.308
{31̄0}⟨130⟩	0.432	0.432	0	0.433	0.209	0.241

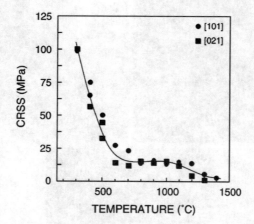

Fig. 5.37 Temperature dependence of CRSS for slip on (001)⟨110⟩ in TiSi$_2$. Inui et al. (2003). With kind permission of Elsevier

The dislocations gliding on (001) are dissociated according the 1/2[110] scheme are shown in Fig. 5.39 (the same thin foil is used as in Fig. 5.38). The coupled partial dislocations seen in Fig. 5.39a are invisible for $\mathbf{g} = \bar{3}3\bar{1}$ (Fig. 5.39b) and $\mathbf{g} = \bar{3}31$ (Fig. 5.39c). It indicates that both partial dislocations have Burgers vector $\mathbf{b} = 1/4[110]$. The dissociation is as follows:

$$1/2\langle 110 \rangle \rightarrow 1/4\langle 110 \rangle + SF + 1/4\langle 110 \rangle \tag{5.15}$$

Clearly SF stands for stacking fault. The stacking fault energy is 164 mJ m^{-2}.

The dislocation structures of deformed TiSi$_2$ crystals with orientation [101] at the temperatures of 400 °C, 1000 °C and 1300 °C are illustrated in the respective shown in Fig. 5.40. The Burgers vector is of 1/2[110] type.

At the temperature where the stress-strain curves showed serrations, the dislocations observed were wavy as indicated by the arrows in Fig. 5.40b. At temperatures above 1200 °C there is a tendency for the dislocations to be out of their slip planes suggesting climb contribution to deformation. Some of the dislocations with two different ⟨110⟩ Burgers vector at the high temperatures tend to form nodes (double arrows in Fig. 5.40c) according to

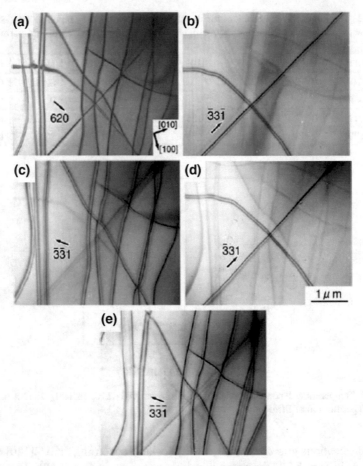

Fig. 5.38 Contrast analysis of dislocations introduced in a TiSi$_2$ single crystal with the [101] orientation deformed at 500 °C. The thin foil was cut parallel to (001) slip planes and the specimen geometry is indicated in (**a**). The operating reflections are indicated in the figures. Inui et al. (2003). With kind permission of Elsevier

$$1/2[110] + 1/2[\bar{1}10] \rightarrow [010] \qquad\qquad (5.16)$$

Weak beam images of 1/2[110] dislocations are seen—at the temperatures indicated above—in Fig. 5.41. Note the dislocation constrictions in 5.41b, c. It is believed that the restriction seen at the higher temperature are the result of dislocations-vacancies reaction. It occurs in the temperature range of 800–1100 °C, the temperature range where serrated flow stress is observed. Therefore serration in the flow stress is associated with dislocations-vacancies interactions.

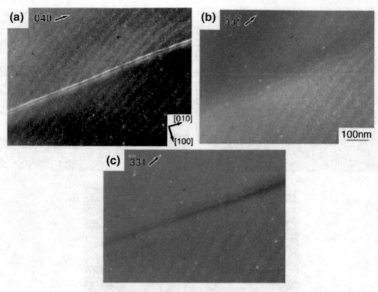

Fig. 5.39 Weak-beam contrast analysis of 1/2[110] dislocations in a TiSi₂ single crystal with the [101] orientation deformed at 500 °C. The thin foil was cut parallel to (001) slip planes and the operating reflections are indicated in the figures. Inui et al. (2003). With kind permission of Elsevier

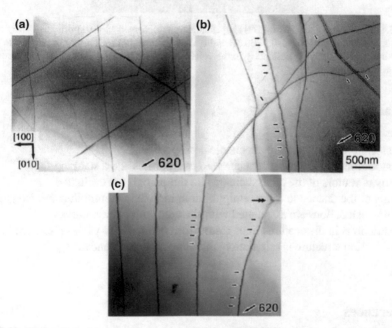

Fig. 5.40 Dislocation structures observed in [101]-oriented TiSi₂ single crystals deformed at **a** 400 °C, **b** 1000 °C and **c** 1300 °C. All thin foils were cut parallel to (001) slip planes and the specimen geometry for all foils is indicated in (**a**). Inui et al. (2003). With kind permission of Elsevier

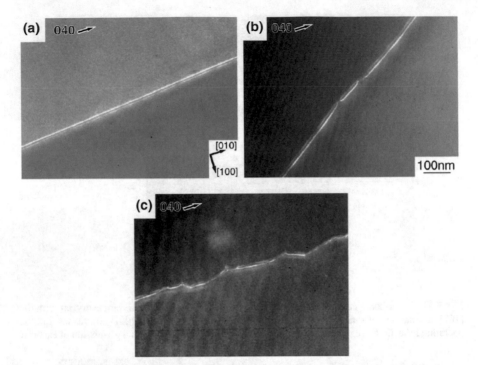

Fig. 5.41 Weak-beam images of 1/2[110] dislocations observed in [101]-oriented TiSi$_2$ single crystals deformed at **a** 400 °C, **b** 1000 °C and **c** 1300 °C. All thin foils were cut parallel to (001) slip planes and the specimen geometry for all foils is indicated in (**a**). Inui et al. (2003). With kind permission of Elsevier

Summary

- Dislocations determine the mechanical properties
- Dislocations in silicides dissociate into two partials and a stacking fault
- Burgers vectors of the undissociated and the partials are evaluated
- Many of the dislocations are straight bur at high temperature they are irregular
- Wavy dislocations are associated with serrated stress-strain curves
- Restrictions in dislocations lines result from dislocations-vacancies reaction
- Dislocation structure is orientation and temperature dependent.

References

P.N.B. Anongba, S.G. Steinemann, Phys. Stat. Sol. **140**, 391 (1993)
B. Bokhonov, M. Korchagin, J. Alloys and Compounds **319**, 187 (2001)
L.J. Chen, J.W. Mayer, K.N. Tu, Thin Solid Films **93**, 135 (1982)
A.H. Cottrell, Phil. Mag. **74**, 829 (1953)

C. Dietzsch, M. Bartsch, and U. Messerschmidt, Phys. Stat. Sol. (a), **202**, 2249 (2005)
J.M. Gibson, J.C. Bean, J.M. Poate, R.T. Tung, Thin Solid Films **93**, 99 (1982)
H. Inui, M. Moriwaki, N. Okamoto, M. Yamaguchi, Acta Mater. **51**, 1409 (2003)
K. Ito, H. Inui, T. Hirano, M. Yamaguchi, Acta Metall. Mater. **42**, 1261 (1994)
K. Ito, T. Yano, T. Nakamoto, H. Inui, M. Yamaguchi, Acta Mater. **47**, 837 (1999)
E. Orowan, Z. Phys. 605, 614, 634 (1934)
J. Pelleg, *Mechanical Properties of Materials* (Springer, Dordrecht, 2013)
J. Pelleg, *Mechanical Properties of Ceramics* (Springer, 2014)
M. Polanyi, Z. Phys. 605, 614, 634 (1934)
G.I. Taylor, Proc. R. Soc. **145**, 362 (1934)
R.T. Tung, J.M. Poate, J.C. Bean, J.M. Gibson, D.C. Jacobson, Thin Solid Films **93**, 77 (1982)

Chapter 6
Time Dependent Deformation—Creep in Silicides

Abstract Andrade was the first to formulate creep phenomena indicating its stress, time and temperature dependence. Basic conditions for β and κ creeps are presented. Disregarding instant elongation, creep is known to occur at three stages: transient creep, steady state creep and accelerated creep. Limited information on creep in silicides exist and therefore only $MoSi_2$, $MoSi_2$-WSi_2 and $TiSi_2$ are considered. Grain size has an important effect on the creep behavior and in polycrystals large grain size is essential, but single crystalline components are preferential. Basic relations for creep are presented for the creep rate showing the importance of d the grain size, p the grain size exponent and n the stress exponent. There is a transition from Newtonian viscous flow which is self diffusion dependent to a power-law creep associated with dislocation climb. Dislocation creep involves glide and climb and both are associated with diffusion and the slowest of them controls the creep rate. A threshold stress and temperature exist below which creep will not occur.

6.1 Fundamentals of Creep

In this section a brief summary is presented on creep. Detailed discussion on creep can be found in the book on "Creep in Ceramics" (Pelleg 2017) and Mechanical Properties of Materials (Pelleg 2017). Andrade was the first to formulate creep, expressing creep rate as a function of stress, time and temperature, namely

$$\dot{\varepsilon} = f(\sigma, t, T) \tag{6.1}$$

Most of the equations following Andrade's concept express creep strain by the empirical relation (Andrade 1910, 1956)

$$\varepsilon = \varepsilon_0\left(1 + \beta t^{1/3}\right)\exp(\kappa t) \tag{6.2}$$

When $\kappa = 0$, Eq. (6.2) reduces to

$$\varepsilon = \varepsilon_0\left(1 + \beta t^{1/3}\right) \tag{6.3}$$

© Springer Nature Switzerland AG 2019   107
J. Pelleg, *Mechanical Properties of Silicon Based Compounds: Silicides*,
Engineering Materials, https://doi.org/10.1007/978-3-030-22598-8_6

Relation (6.3) is known as β creep. Equation (6.3) represents transient creep, since creep is decreasing over time. When however β = 0, Eq. (6.2) is written as

$$\varepsilon = \varepsilon_0 \exp(\kappa t) \qquad (6.4)$$

Equation (6.4) is known as the κ creep and represents a stationary state. Differentiating Eqs. (6.3) and (6.4) results in

$$\frac{d\varepsilon}{dt} = \dot{\varepsilon} = \frac{1}{3}\varepsilon_0 \beta t^{-2/3} \qquad (6.5)$$

and

$$\dot{\varepsilon} = \kappa\varepsilon_0 exp(\kappa t) = \kappa\varepsilon \qquad (6.6)$$

Recapturing schematically the familiar creep curves along the lines of Andrade they are illustrated in Fig. 6.1 showing three stages of creep and an instantaneous elongation on application of the load.

Three stages of creep are seen in Fig. 6.1 when the plotting is done at constant load. In the first stage creep-also known as transient creep or primary creep—strain hardening occurs which slowly declines. In the second stage creep—also known as steady state or linear creep-the strain rate is constant (Fig. 6.1b). The constant creep rate is the minimum creep rate which is an important design parameter and its magnitude is temperature and stress dependent. The third stage—also known as the tertiary or accelerated creep-the creep progress is accelerated up to fracture.

Fig. 6.1 a A schematic creep curve along the lines of Andrade. A constant stress curve is incorporated in the overall illustration indicated by the dashed line extension; **b** schematic strain rate plot versus time; Pelleg (2013)

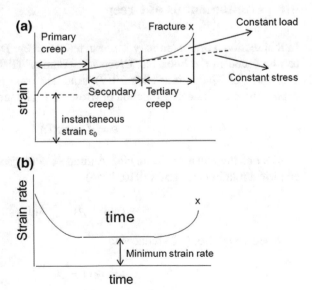

The interest is to avoid all forms of creep while a component is exposed to some temperature. Therefore, it is important to evaluate a threshold stress and temperature—the additional factor of time- below which creep will not occur. Norton attempted to determine this threshold by suggesting a relation based on the observation that a constant stress produces a constant secondary creep rate as

$$\dot{\varepsilon} = A\sigma^n \tag{6.7}$$

In this relation A and n are constants determined experimentally that are functions of temperature only.

6.2 Creep in CoSi$_2$

No direct information on time dependent deformation is available in the literature. However a comparison of various silicides in the range of 1000–1400 °C contains information on the minimum creep rate of CoSi$_2$ also. In Fig. 6.1 minimum creep rate is plotted versus stress. It is immediately obvious that CoSi$_2$ shows the least creep resistance among the alloys illustrated in Fig. 6.2. In particular note the creep of MoSi$_2$ which is better than that of CoSi$_2$ and has been widely studied for its creep properties. The reason probably is the better average mechanical properties as a function of temperature and its higher melting point than that of CoSi$_2$.

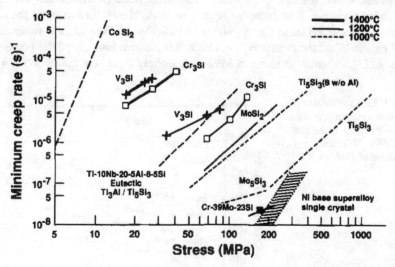

Fig. 6.2 Comparison of minimum creep rate in compression versus stress for several silicides and their alloys. Shah et al. (1992). With kind permission of Elsevier

6.3   Creep in MoSi$_2$

For application at temperatures above 1000 °C, MoSi$_2$ received considerable interest among the silicides, due to its relatively good oxidation resistance and good creep properties. The tests performed were under compression. The MoSi$_2$ specimens were polycrystalline. Hence the grain size has an important effect on the creep behavior. For example grain boundary sliding might be involved in components exposed to creep. An additional equation to those of (6.2)–(6.7) is presented in Eq. 6.8 which takes specifically into account the grain size effect. It is the power-law relationship.

$$\dot{\varepsilon} = A(1/d)^p(\sigma)^n \exp(\Delta Q/RT) \tag{6.8}$$

A is a structure factor, d is the grain size, p is the grain size exponent, n is the stress exponent and ΔQ is the activation enthalpy. In Fig. 6.3 the creep rate is plotted versus stress for several grain sizes at 1200 °C for a stress exponent n = 1. The well known fact that larger grain sized material creep at a smaller rate can be observed in Fig. 6.3 also, where it is seen that at a given stress the large grain size MoSi$_2$ creeps at a small rate. One could deduce form these observations, that in designing a specific material for creep resistance large grain material should be chosen. The creep rate variation with the inverse grain size at 1200 °C with a grain size exponent p = 4.3 and stress exponent n = 1 is shown in Fig. 6.4.

In plotting Fig. 6.4 (according to Eq. 6.8) it is assumed that all factors except the grain size are independent of the grain size which is not a realistic possibility.

Considering two stages of creep such as indicated in Fig. 6.1 when loading is at constant stress, the first stage creep rate varies linearly with stress indicating a Newtonian viscous flow behavior with n = 1. As the stress increases the creep occurs in the second stage (steady state) as indicated in Fig. 6.1, the value of the stress exponent increases to n = 3–4. Often this creep is also called the power-law creep. At higher stress, in terms of load—the tertiary creep, the value of n > 4 and

Fig. 6.3 Effect of stress and grain size on creep rate in the Newtonian creep. Sadananda et al. (1999). With kind permission of Elsevier

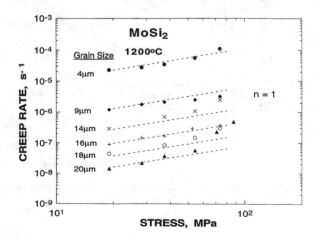

Fig. 6.4 Creep rates as a
function of inverse grain size
from Fig. 6.3. Sadananda
(1999). With kind
permission of Elsevier

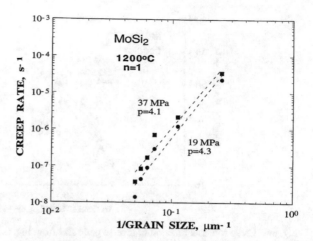

deviations from the power law are observed. In Fig. 6.5 the effect of grain size is
seen for creep at higher stress at 1200 °C and at n = 4. As mentioned earlier for
creep design high grain material is preferable because the creep rate is smaller. This
is clearly seen in the case of creep in MoSi$_2$ as indicated in Fig. 6.5. The decrease
in grain size to 4 μm much increases the creep rate by a few orders compared to
the larger 25 μm grain sized MoSi$_2$. Apparently at the large grain sized MoSi$_2$ the
two points at the larger stress deviate from linearity, indicating the possibility that
the power law is not applicable at the high stresses. The validity of the power law at
1200 °C for creep specimens with the stress exponent n = 4 and grain size exponent
p = 4.3–4.4 is illustrated in Fig. 6.6 where the creep rate is plotted versus the inverse
grain size. The results of step loading from the Newtonian flow behavior (viscous)
with n = 1 to the power law behavior with n = 4 at 1200 °C is collected for the
various grain sizes in Fig. 6.7. As can be seen the grain size effect is not observed
in the power-law line of the Newtonian preloaded data (precept at the lower stress

Fig. 6.5 Effect of stress and
grain size on creep rate in the
power-law creep. Sadananda
et al. (1999). With kind
permission of Elsevier

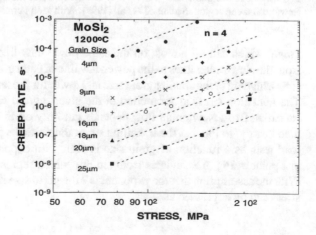

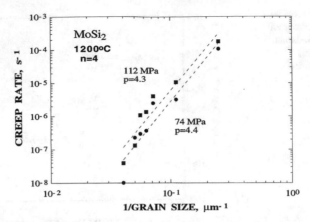

Fig. 6.6 Creep rate as a function of inverse grain size from Fig. 6.5. Sadananda et al. (1999). With kind permission of Elsevier

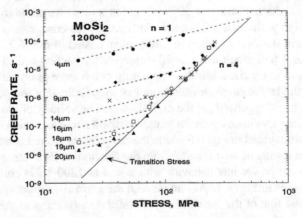

Fig. 6.7 Effect of grain size in the power-law creep when the specimens step loaded from the Newtonian creep region. Sadananda et al. (1999). With kind permission of Elsevier

range) which fall all almost on the same power-law line. The transition, however, from the Newtonian lines to the power-law line of each grain size occurs at a different stress, higher for the smaller grain size and lower for the larger grain size respectively. The transition stress as a function of the inverse grain size is seen in Fig. 6.8. The transition from Newtonian flow depends not only on the grain size but also on the load history. In Fig. 6.8 the transition stress that initiates the power law in increasing load tests as a function of grain size is seen with exponent between 2 and 3. The last point in Fig. 6.8 deviates from the line which represent the smallest grain size. With the assumption that the exponent is 2 the grain size dependence of the transition stress can be expressed as

$$\sigma_T = \sigma_0 + (K_T/d)^2 \qquad (6.9)$$

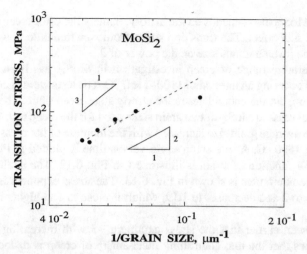

Fig. 6.8 Effect of grain size on the stress for transition from Newtonian to the power-law creep from Fig. 6.7. Sadananda et al. (1999). With kind permission of Elsevier

The activation energy dependence on grain size is shown in Fig. 6.9. At the beginning the activation energy increases for small grain sizes reaching a plateau at grain size $>\sim15$–$20 \ \mu m$ from 180 to 375 kJ mol^{-1}. LANL in Fig. 6.9 indicates the origin of the powder for the MoSi₂ specimens and stands for Los Alamos National Laboratory.

Thus, it is possible to conclude that the behavior of MoSi₂ depends significantly on the grain size even in the power-law creep where n = 4. The grain size exponent p is ~4.3 in both the Newtonian and the power law ranges. This value of the grain size exponent is higher than either the Coble or the Nabarro-Herring creep. The power

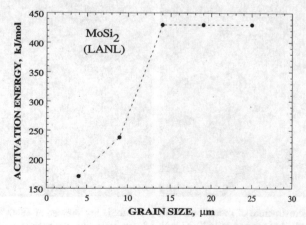

Fig. 6.9 Effect of grain size on the activation energy for creep. Sadananda et al. (1999). With kind permission of Elsevier

law creep in MoSi$_2$ also depends on the history, namely the prior creep deformation and the grain size effect. The transition stress from Newtonian to power law creep depends on the inverse grains size on the power of 2.

The temperature range of creep investigation in MoSi$_2$ has been extended to 1450 °C, thus covering an interval of 1100–1450 °C. The tests were again performed by compression, but the data relates to the steady state creep stage. The microstructures of the hot pressed MoSi$_2$ at two grain sizes used for the creep tests are shown in Fig. 6.10. The average grain size increased with the increase of hot press temperature from 1600 to 1800 °C. A true creep strain versus time is plotted in Fig. 6.11. The creep rate as a function of strain is illustrated in Fig. 6.12. The strain rate versus stress at the temperatures is shown in Fig. 6.13. The stress exponent is about 1.93, namely close to 2 and decreases to 1.19, which is close to 1 at higher temperatures and stresses.

We have seen earlier that the stress exponent ~3 with increasing temperature usually implies that the rate controlling mechanism of creep is dislocation climb. Stress exponent n < 3 correspond to a transition region from Newtonian viscous flow to power-law creep involving dislocation climb and glide. An apparent activation energy for creep in MoSi$_2$ can be determined from Fig. 6.14 where the strain rate is plotted versus the inverse temperature. The value of the activation energy determined is 433 kJ mol^{-1}.

The creep rate can be expressed by one of the creep rate equations taking into account the stress exponent, the temperature and the stress as

$$\dot{\varepsilon} = A\sigma^n \exp(-Q/RT) \tag{6.10}$$

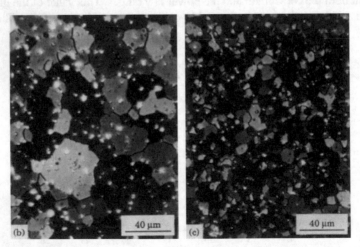

Fig. 6.10 a Microstructure of monolithic MoSi$_2$: **b** MoSi$_2$ hot pressed at 1820 °C with 35 μm grain size; **c** MoSi$_2$ hot pressed at 1620 °C with 18 μm grain size. Sadananda et al. (1992). With kind permission of Elsevier

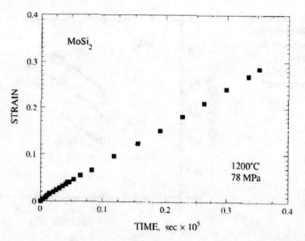

Fig. 6.11 Typical creep curve for monolithic MoSi₂. Sadananda et al. (1992). With kind permission of Elsevier

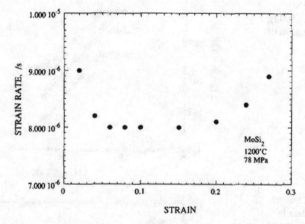

Fig. 6.12 Creep rates as a function of creep strain for monolithic MoSi₂. Sadananda et al. (1992). With kind permission of Elsevier

From a least square analysis of the data A, n and a correlation factor were evaluated. The least square analysis data are presented in Fig. 6.15 and resulting in

$$\dot{\varepsilon} = 4.432 \times 10^4 \sigma^{1.75} \exp(-380/RT) \tag{6.11}$$

The Q value in Eq. 6.11 and Fig. 6.15 representative of the steady state value which is lower than Q determined on the basis of Fig. 6.14. The least square analysis is presented in Table 6.1. TEM investigation was used to show the dislocation structure as illustrated in Fig. 6.16.

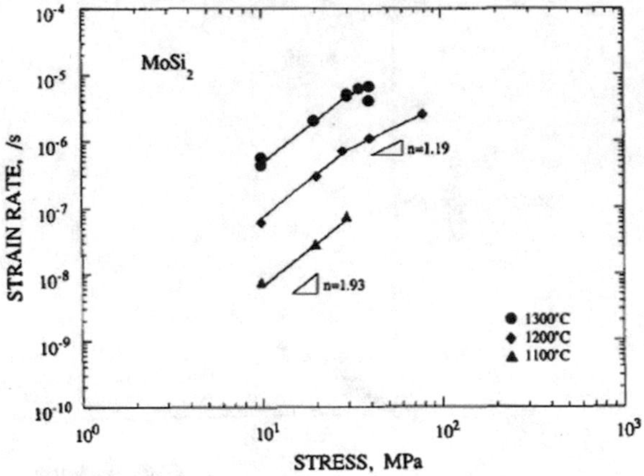

Fig. 6.13 Stady state creep
rates as a function of applied
stress and temperature for
monolithic $MoSi_2$.
Sadananda et al. (1992).
With kind permission of
Elsevier

Fig. 6.14 Arrhenius plot for
the determination of the
apparent activation energy
for creep for $MoSi_2$.
Sadananda et al. (1992).
With kind permission of
Elsevier

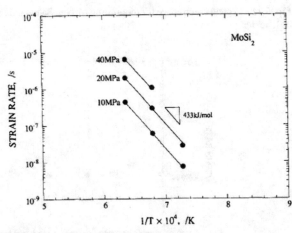

Fig. 6.15 Least square
analysis of the data to
determine constitution
equation for creep in $MoSi_2$.
Sadananda et al. (1992).
With kind permission of
Elsevier

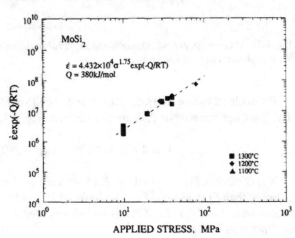

Table 6.1 Least square analysis to determine creep constants Q, A and n for MoSi$_2$. $\dot{\varepsilon}$ = $A\sigma^n\exp(-Q/RT)$, σ is applied stress. Sadananda et al. (1992). With kind permission of Elsevier

Material	Q (kJ mol^{-1})	A	n	r
MoSi$_2$	350	4.029×10^3	1.749	0.97739
	360	8.961×10^3	1.750	0.98034
	370	1.993×10^4	1.750	0.98220
	380	4.432×10^4	1.751	0.98302
	390	9.857×10^4	1.751	0.98286
	400	2.192×10^5	1.751	0.98174
	410	4.875×10^5	1.752	0.97971

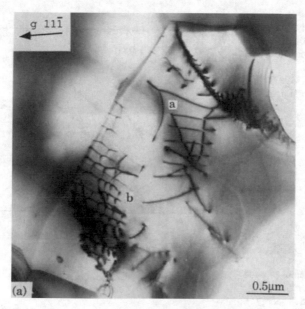

Fig. 6.16 TEM examination of dislocation in the MoSi$_2$ matrix. Sadananda et al. (1992). With kind permission of Elsevier

Thus it was indicated above that there is a transition from Newtonian viscous deformation which is associated with self diffusion to a power-law creep associated with dislocation climb. Dislocation creep involves glide and climb. Dislocation climb requires diffusion of both components of MoSi$_2$ but the slowest of them controls the creep rate.

6.4   Creep in MoSi₂-WSi₂

Since no creep data are recorded in the literature for WSi$_2$ and since it forms a solid solution with MoSi$_2$ it was felt of importance to consider this alloy despite the fact that alloying per se is discussed in a separate chapter. Further, in addition to the extended solid solubility and since the melting point and thus the elastic nodulus of WSi$_2$ is higher than that of MoSi$_2$ it is expected that the creep resistance of the alloy will also be higher. Figure 6.17 shows the creep rates in MoSi$_2$-WSi$_2$ alloy as a function of stress and temperature. An apparent activation energy is determined from a plot of strain rate versus the inverse temperature as shown in Fig. 6.18. The activation energy is 536 kJ mol^{-1}.

This value is quiet higher than that of pure MoSi$_2$ by about 100 kJ mol^{-1}. The expectation that the creep resistance of the alloy will be higher was born out by

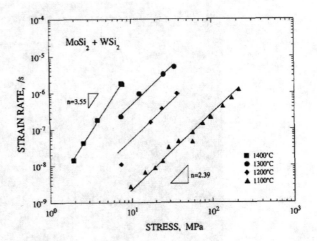

Fig. 6.17 Steady state creep rates as a function of applied stress and temperature for MoSi$_2$-WSi$_2$ alloy. Sadananda et al. (1992). With kind permission of Elsevier

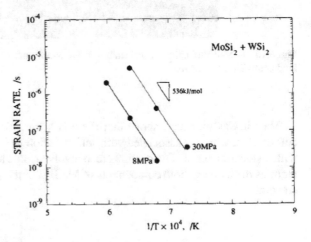

Fig. 6.18 Arrhenius plot for the determination of apparent activation energy for creep in MoSi$_2$-WSi$_2$ alloy. Sadananda et al. (1992). With kind permission of Elsevier

Fig. 6.19 Effect of alloying on the creep resistance of monolithic MoSi₂. Sadananda et al. (1992). With kind permission of Elsevier

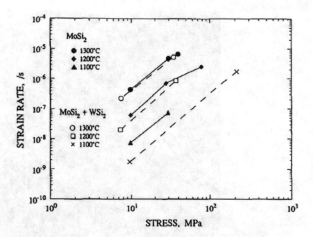

experimental results as seen in Fig. 6.19 where the strain rate is plotted versus stress. Note the creep resistance expressed as strain rate at the lower temperatures is better in MoSi₂-WSi₂ than in MoSi₂. However this improvement decreases with increasing temperature and already at 1300 °C the strain rates of the two alloys almost coincide at each applied stress.

6.5   Creep in TiSi₂

As we have indicated earlier regarding CoSi₂ and NiSi₂ similarly almost no information on the creep properties of TiSi2 exists. Considerable information however exist on their applications in electronic thin-film devices since this is the primary interest in these compounds. The limited data on creep in TiSi₂ is presented below. The TEM bright field micrograph of the dislocation structure in undeformed TiSi₂ is illustrated in Fig. 6.20. The dislocation structure of the creep deformed TiSi₂ is shown in Fig. 6.21. The TiSi₂ samples were creep tested in compression in the temperature range 800–1200 °C in air at constant strain rates. As indicated earlier no other creep data have been reported in the literature. The creep can be described by the power law as indicated in earlier parts of this chapter by (Table 6.2)

$$\dot{\varepsilon} = A \frac{D_L G \left| \vec{b} \right|}{kT} \left(\frac{\sigma}{G} \right)^n \tag{6.12}$$

The stress exponent n = 2.8 ± 0.2 predicts a power law creep behavior by a viscous glide of dislocations supported by dislocation mechanism. The stress exponent is calculated from the slope from the strain rate versus creep stress curves shown in Fig. 6.22. The creep stress for TiSi₂ in the range 900–1000 °C was determined to 56 or 17 MPa at a strain rate of $\dot{\varepsilon} = 10^{-7}\,\mathrm{s}^{-1}$. The major rate controlling process is thermally activated dislocation glide and jogs of screw dislocations segments.

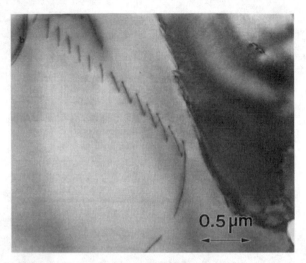

Fig. 6.20 TEM bright field micrograph of the dislocation structure in undeformed TiSi$_2$. Rosenkranz et al. (1992). With kind permission of Elsevier

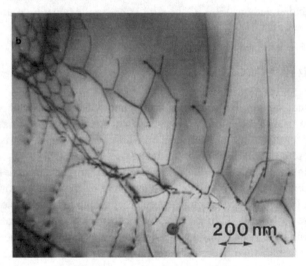

Fig. 6.21 Dislocation structure of creep deformed monolithic TiSi$_2$ compound. ($\dot{\epsilon} = 10^{-7}$ s^{-1}, T = 900 °C). Rosenkranz et al. (1992). With kind permission of Elsevier

Small contribution might come from grain boundary sliding to the overall creep. The activation energy is calculated by the Arrhenius relation given as

$$Q = R\left(\frac{\partial \log \dot{\epsilon}}{\partial(1/T)}\right)_\sigma \tag{6.13}$$

The value Q for TiSi$_2$ is determined to be 320 ± 20 kJ mol^{-1}.

Table 6.2 Least square analysis to determine creep constants Q, A and n for MoSi$_2$-WSi$_2$, $\dot{\varepsilon}$ = $A\sigma^n \exp(-Q/RT)$, σ is applied stress. Sadananda et al. (1992). With kind permission of Elsevier

Material	Q (kJ mol^{-1})	A	n	r
MoSi$_2$-WSi$_2$	520	5.491 × 10^8	2.206	0.99877
	540	2.236 × 10^9	2.274	0.99907
	560	9.101 × 10^9	2.341	0.99919
	580	3.705 × 10^{10}	2.408	0.99916
	610	3.043 × 10^{11}	2.509	0.99887

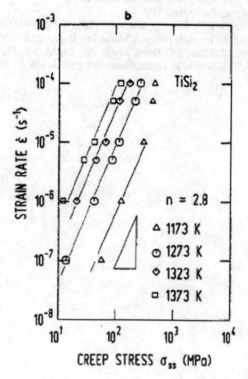

Fig. 6.22 Determination of the stress exponent n from the slope log $\dot{\varepsilon}$ versus log σ plot for TiSi$_2$. Rosenkranz et al. (1992). With kind permission of Elsevier

Summary

- Creep consists of three stages
- Grain size is a significant factor to improve creep resistance
- The stress exponent, p, the grain size exponent, n determine the creep characteristics

- Both Newtonian viscous flow and power law creep are associated with self diffusion
- Dislocation glide and climb, both diffusion controlled determine the creep rate.

References

A.N.daC. Andrade, Proc. R. Soc. Lond. A **84**, 1 (1910)
A.N.daC. Andrade, in *ASM, Creep and Recovery* (American Society of Metals, 1956), p. 176
J. Pelleg, *Mechanical Properties of Materials* (Springer, 2013)
J. Pelleg, *Creep in Ceramics* (Springer, 2017)
R. Rosenkranz, G. Frommeyer, W. Smarsly, Mater. Sci. Eng. A **152**, 288 (1992)
K. Sadananda, C.R. Feng, H. Jones, J. Petrovic, Mater. Sci. Eng. A **155**, 227 (1992)
K. Sadananda, C.R. Feng, R. Mitra, S.C. Deevi, Mater. Sci. Eng. A **261**, 223 (1999)
D.M. Shah, D. Berczik, D.L. Anton, R. Hecht, Mater. Sci. Eng. A **155**, 45 (1992)

Chapter 7
Fatigue in Silicide Composites

Abstract Repeated loading cyclically or a fluctuating stress may induce fatigue in a component. Plots of S-N curves define an endurance limit which in Fe and Ti is a horizontal line. At the stress level defined by the horizontal line or below it a material endures a very large number of stress cycles without failure. However, in most other materials no definite endurance limit exists. The term "runout" refers to about 10^7 cycles representing the highest stress of "non-failure". Avoiding catastrophic and unexpected failure requires the best possible design and choice of material, therefore various combinations of components rather than the pure components are used to avoid fatigue faiure. These reinforcements may be other silicides, aluminides, carbides, nitrides or elements. $MoSi_2$ reinforced with Nb in various forms is a most commonly used composite silicide and the largest improvement occurs when it is added as fibers. Basic fatigue equations are presented. The law of Paris' is included which relates fatigue crack growth to the stress intensity factor.

7.1 Basics

The majority of failure in materials occurs by fatigue. As a matter of fact it is generally believed that over 80% of all service failures are associated with fatigue. Fatigue may occur when a component is subjected to repeated loading cyclically or due to the action of a fluctuating stress. The fatigue failure is manifested by the development of a crack, its growth and propagation till fracture sets in. The force applied may act axially, torsionally or flexurally inducing fatigue failure. A structure exposed to repeated loading can undergo a progressive damage and the danger in fatigue failure is that it might occur without any warning catastrophically at stress levels much below the static yield stress. The failure is always a brittle fracture regardless of whether the material is brittle or ductile. Due to the wide spread of test results, often even a deviation of ~50% from the average value is observed, many test specimens are used in fatigue experiments to reach a meaningful average value. The test results are collected into S-N curves which are plots of the repeated loading stress (S) versus the number of cycles (N) and define the term endurance limit. Fe and Ti, for example show a definite endurance limit (a horizontal line) the meaning of which is that at

© Springer Nature Switzerland AG 2019
J. Pelleg, *Mechanical Properties of Silicon Based Compounds: Silicides*,
Engineering Materials, https://doi.org/10.1007/978-3-030-22598-8_7

this stress level (and below it) the material can endure a very large number of stress cycles without failure (actually never fails). However, in most other materials no definite endurance limit exists and therefore a definite number of cycles is defined arbitrarily that they can endure without failure. In the low cycle fatigue the number of cycles $N < 10^5$ (high loads elastic and plastic deformation) while in the high cycle fatigue $N > 10^5$ (low loads elastic deformation). The term "runout" refers to at least 10^7 cycles representing the highest stress of "non-failure". Safe use in practice is considered when the test is terminated at ~10^8 cycles without failure. In Fig, 7.1 an S-N curve is shown schematically for a material with and without endurance limit, respectively.

Various wave forms of cyclic stresses may be applied to a specimen exposed to fatigue testing in order to evaluate S-N plots. Figure 7.2 shows the various wave

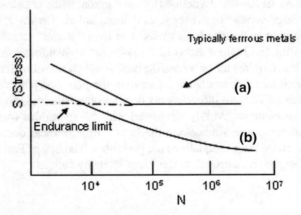

Fig. 7.1 S-N curves: **a** with a well-defined endurance limit, **b** without a definite fatigue limit (typically Al alloys). Pelleg (2014)

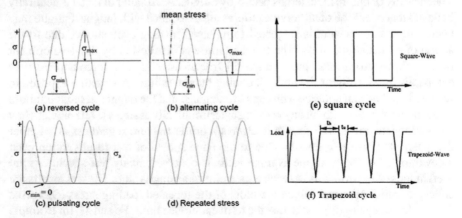

Fig. 7.2 Various forms of cyclic stresses: **a–c** are sinusoidally varying cycles **d** represents a repeating cycle, **e** a square cycle and **f** a trapezoid cycle. Pelleg (2014)

forms of the cyclic stresses that may be applied to a specimen during fatigue testing simulating the real possibilities of a component exposed to fatigue.

At this point it might be useful to indicate some of the basic fatigue equations:

$$\text{Stress range:} \quad \Delta\sigma = \sigma_{max} - \sigma_{min} \tag{7.1}$$

$$\text{Mean stress:} \quad \sigma_m = \frac{\sigma_{max} + \sigma_{min}}{2} \tag{7.2}$$

$$\text{alternating stress:} \quad \sigma_a = \frac{\Delta\sigma}{2} \tag{7.3}$$

$$\text{stress ratio:} \quad R = \frac{\sigma_{min}}{\sigma_{max}} \tag{7.4}$$

$$\text{amplitude ratio:} \quad A = \frac{\sigma_a}{\sigma_m} \tag{7.5}$$

For more details on fatigue, the reader is referred to earlier works by the author of this book.

7.2 Silicide Composites

It is clear that the use of components for avoiding catastrophic and unexpected failure requires the best possible design and choice of material for the given purpose. Due to this requirement almost all if not all effort is concentrated to that goal and therefore information in the literature considers mainly various combinations of components rather than the pure or monolithic silicides. Only the most commonly used or investigated composite silicide is considered below.

7.2.1 Fatigue in MoSi₂ Composite

$MoSi_2$ is a potential material for high temperature structural applications primarily due to its high melting point (2020 °C) and relatively low density (6.3 g/cm³). Further the considerable interest in $MoSi_2$ its relatively good oxidation resistance. By suitable reinforcements, significant enhancement in strength can be achieved and its reasonable ductility makes it convenient for EDM (Electrical discharge machining). The strengthening additives to the $MoSi_2$ may be other silicides ($CrSi_2$, WSi_2, etc.), aluminides (NiAl), carbides (SiC, TiC, etc.), nitrides (Si_3N_4) or elements (Nb). Further the additives may be introduced as particles, laminates, wires etc.

7.2.1.1 Nb Reinforcement

The composites were prepared from $MoSi_2$ powder reinforced with Nb powder consolidated by hot isostatic pressing (HIP) under 207 MPa at 1700 °C for 4 h. Particulate, fiber and laminate Nb reinforcement are considered below. The uniform distribution of the Nb particulate in the $MoSi_2$ matrix is shown in Fig. 7.3. Fiber reinforcement is illustrated in Fig. 7.4 and laminated structure of $MoSi_2$/Nb is seen in Fig. 7.5. All the composites contained 20% Nb by volume. Pure Nb powder was consolidated by the same processing conditions as the composite for comparing the fatigue properties.

The fatigue tests were performed on single edge notched (SEN) specimens. The initial noth-to-width ratio was 0.25. The fatigue crack growth test specimens were precracked and then subjected to a constant amplitude cyclic loading at a stress ratio $R = K_{min}/K_{max} = 0.1$. The cyclic loading frequency of 10 Hz was applied. In the above K is the stress intensity factor. The low initial stress ranges $\Delta\sigma = \sigma_{max} - \sigma_{min}$, corresponding to a stress intensity range of 1 MPa $\sqrt{m}$ were applied initially for incremental loading stages of 10^6 cycles. The stress ranges were increased in incremental stage of 10% until crack growth was detected after 10^6 cycles. On detection of the crack growth the stress amplitudes were maintained henceforth constant. For the results of reinforcement in static experiments under monotonic loading the reader is referred to the work of Soboyejo et al. (1996).

The fatigue crack growth rate in the composite and the pure Nb is presented in Fig. 7.6. The fatigue crack growth in the composite was faster than in pure $MoSi_2$, and the Paris' exponent was also high ~9.4 compared to the typically observed 2–4 in monolithic metals. Recall that the Paris' law relates the fatigue crack growth to the stress intensity factor range according to

$$\frac{da}{dN} = C\Delta K^m \tag{7.6}$$

where a is the crack length da/dN is the crack growth rate denoting the infinitesimal grack length growth per increasing number of load cycles. C and ΔK are constants

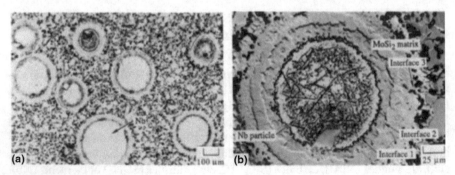

Fig. 7.3 Microstructure of particulate $MoSi_2$/Nb composites: **a** particle distribution; and **b** particle/matrix interface. Soboyejo et al. (1996). With kind permission of Elsevier

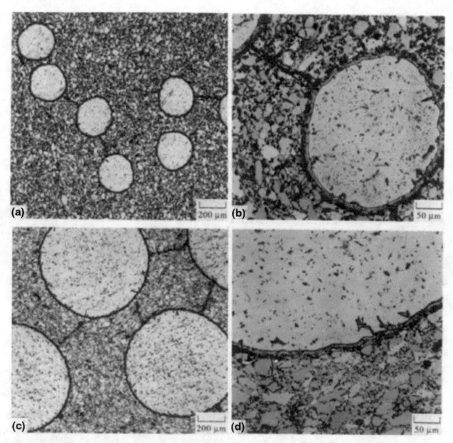

Fig. 7.4 Microstructure of fiber-reinforced MoSi$_2$/Nb composites: **a, b** 250 μm diameter fiber; and **c, d** 750 μm diameter fiber. Soboyejo et al. (1996). With kind permission of Elsevier

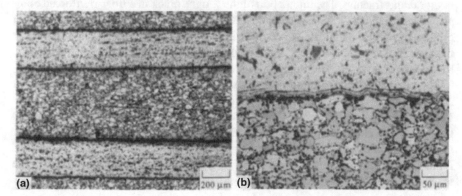

Fig. 7.5 Microstructures of laminated MoSi$_2$/Nb composites: **a** laminate distribution; and **b** laminate/matrix interface. Soboyejo et al. (1996). With kind permission of Elsevier

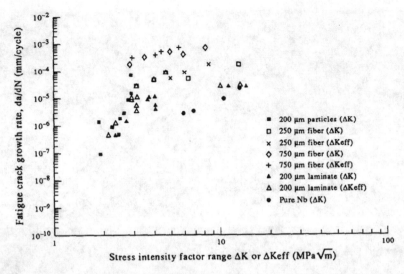

Fig. 7.6 Summary of fatigue crack growth rate data. Soboyejo et al. (1996). With kind permission of Elsevier

that depend on the material, environment and stress ratio. ΔK the range of the intensity factor is given by

$$\Delta K = K_{max} - K_{min} \tag{7.7}$$

In the composite the crack tends to deflect around the interfaces of the Nb, whereas in the monolithic $MoSi_2$ the path of the crack is relatively straight. The crack path observed is shown in Fig. 7.8. Fatigue crack growth in th $MoSi_2$ matrix and in the composite is shown for various cases in Fig. 7.7. The fatigue crack growth in the matrix seen in Fig. 7.8a, b occurred mainly by transgranular cleavage and some intergranular fracture. The interaction of the fatigue crack in a fiber-reinforced composite is seen in Fig. 7.8c. The damage initiated from the notch. No evidence of fiber rupture was observed until ΔK reached high values. The fracture mechanism in the fiber reinforced composite (Fig. 7.8c, d) and in the particulate composite (Fig. 7.8a, b) are similar. The Nb fiber failed by cleavage under cyclic loading.

It was observed that fatigue crack growth rate in composites with 750 μm fiber addition was faster than those reinforced by a smaller diameter Nb wire (250 μm). The crack growth was also faster than in the Nb laminated composites with 200 μm thickness. The composite containing particulate reinforcement had about the same crack growth rate as the laminated one. Finally the crack growth rate in the composites is much faster than in the monolithic Nb. Principally the crack growth in the composite is intermittent rather than continuous.

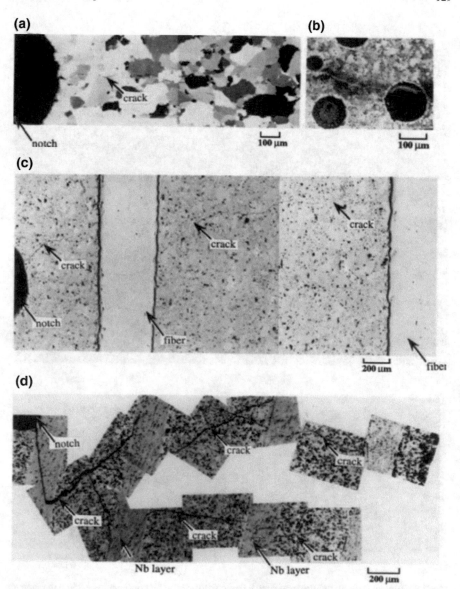

Fig. 7.7 Interactions of fatigue cracks with microstructure: **a** compression pre-crack in monolithic MoSi₂; **b** deflection in particulate-reinforced composites; **c** bridging and blunting in fiber composite; and **d** multiple cracks in laminated composite. Soboyejo et al. (1996). With kind permission of Elsevier

7.2.1.2 Reinforcement-Nb Spheres

MoSi₂ powder mixed with the reinforcing particles of Nb spheres ($-35 + 80$ mesh or ($-500 + 177$ μm) were HIP consolidated at 1700 °C for 1 h under 200 MPa in

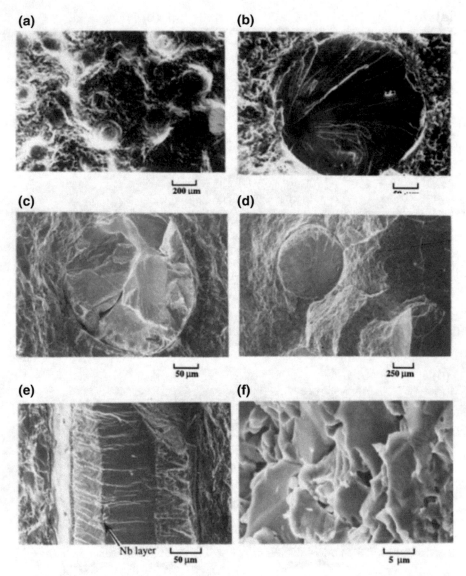

Fig. 7.8 Typical fatigue fracture modes: **a, b** particulate reinforced composites; **c** 250 μm diameter fiber composite; **d** 750 μm diameter fiber; and **e, f** laminated composite. Soboyejo et al. (1996). With kind permission of Elsevier

argon. The nominal reinforcing agent was 20% by volume. The $MoSi_2$ matrix was fine grained of ~14 μm. Considerable reaction was observed between the $MoSi_2$ matrix and the Nb resulting in ~40 μm reaction layer around the Nb particles. Tests were performed at cyclic loading at a load ratio $R = K_{min}/K_{max} = 0.1$ and a frequency of 25 Hz. Crack length was continuously monitored in situ. The results are presented as the growth rate per cycle da/dN versus the applies stress sensitivity range ΔK_{sp} (= K_{max} = K_{min}). The fatigue crack growth is presented in Fig. 7.9. Fatigue crack growth is illustrated in Fig. 7.10.

As can be seen the growth rate depends quite strongly on ΔK, but the figure is a nonsigmoidal curve. In terms of Paris' power-law (da/dn vs. ΔK) the exponent is ~14 compared to the value 2–4 in metals and 15–50 for ceramics.

The crack-particle interactions in the composite under cyclic loading is not too much different than that observed under monotonic loading in static experiments. Generally the fatigue cracks tend to avoid the ductile Nb reinforcement (see Fig. 7.9a, b). The cracking in the matrix occurs by brittle intergranular and transgranular fracture similar to that in static failure (see Fig. 7.11). However the failure in the Nb particles when encounters crack is by brittle transgranular-shear (see Fig. 7.12) rather than transgranular cleavage as observed in monotonic loading. Only moderate improvement in fracture toughness values were found in the composite compared to the unreinforced $MoSi_2$ matrix material.

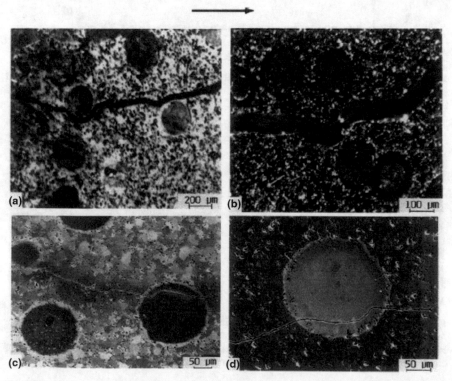

Fig. 7.9 **a** and **b** In situ optical telescope and **c** and **d** SEM images of crack path morphologies in the Nb/$MoSi_2$ composite, showing crack propagation (**a**) around (**b**) between the Nb spheres. Crack-path trajectories around the Nb generally occur in the (NbMo)$_5$Si$_3$ layer close to matrix/reaction-layer interface. Features are similar under monotonic and cyclic loading. Arrow indicates direction of crack growth. Venkateswara Rao et al. (1992). With kind permission of Springer

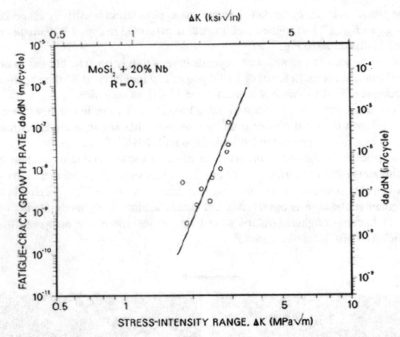

Fig. 7.10 Variation in cyclic fatigue-crack propagation rates, da/dN, as a function of the applied stress-intensity range, ΔK, in the Nb/MoSi$_2$ composite, in controlled room-temperature air at a load ratio R of 0.1. Venkateswara Rao et al. (1992). With kind permission of Springer

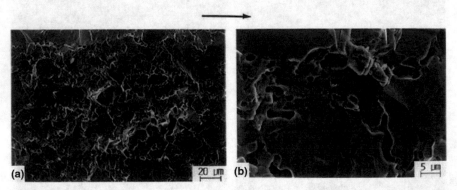

Fig. 7.11 a Low- and **b** high-magnification SEM micrographs of fracture in Nb/MoSi$_2$, under monotonic loading, showing transgranular plus intergranular cracking in the porous MoSi$_2$ matrix. Features are nominally similar under monotonic and cyclic loading. Arrow indicates direction of crack growth. Venkateswara Rao et al. (1992). With kind permission of Springer

7.2.1.3 Reinforcement-Nb Fibers

Crack propagation in cyclic deformation MoSi$_2$ reinforced with 20 vol.% Nb fibers improves greatly the fatigue resistance due to deflection of the crack path. This

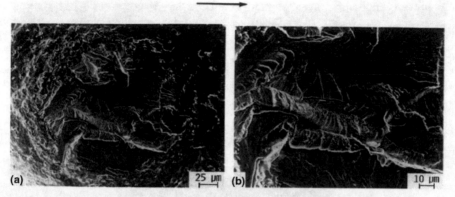

Fig. 7.12 **a** Low- and **b** high magnification SEM micrographs of fracture in Nb/Mosi$_2$ under cyclic loading, showing brittle-shear cracking in the few Nb particles that are intercepted by the crack. Arrow indicates direction of crack advance. Venkateswara Rao et al. (1992). With kind permission of Springer

beneficial effect is associated with crack deflection along the Nb/matrix reaction layer. The crack/Nb wire interaction layer is (Nb, Mo)$_5$Si$_3$ and Fig. 7.13 illustrates the results of the reaction. The variation of the cyclic fatigue crack growth rate is illustrated in Fig. 7.14 at R = 0.1. The plot seen in the figure is of da/dN versus ΔK and is compared with Nb sphere reinforced MoSi$_2$. Results can be expressed in terms of Paris' power-law expressed earlier and rewritten here as

$$da/dN = C\Delta K^m \tag{7.6}$$

In Eq. 7.6, C and m are constants. The exponent is very high in the case of Nb$_m$/MoSi$_2$ of ~20 and in an earlier work a value of Nb$_p$/MoSi$_2$ ~14 was indicated. The crack growth rate in the Nb wire reinforced MoSi$_2$ is orders of magnitude slower than in the Nb particulate (Nb$_p$) reinforced MoSi$_2$. Thus, the fatigue crack-growth threshold, ΔK$_{TH}$ estimated at a minimum growth rate of 10^{-11}m/cycle is 7.2 MPa $\sqrt{m}$ in Nb$_m$/MoSi$_2$ compared to ~2 MPa $\sqrt{m}$ in Nb$_p$/MoSi$_2$.

Fatigue fracture surfaces are seen in Fig. 7.15. The fatigue crack growth rate as a function of the applied intensity range, ΔK at various load ratios is illustrated in Fig. 7.16. A regression analysis to fit the data to the lines in Fig. 7.16 provided the results of

$$da/dN = 6.31 \times 10^{-25}(K_{max})^{13.2}(\Delta K)^{7.5} \tag{7.8}$$

Unlike in Ni based superalloys, the exponent p (of ΔK) in Eq. 7.8 is smaller than the exponent n (of K$_{max}$). In Fig. 7.17 a similar plot is illustrated but as a function of K$_{max}$ rather than ΔK.

Much better improvement in fatigue resistance as indicated by crack growth was obtained in Nb wire reinforced MoSi$_2$ (Nb$_m$/MoSi$_2$) than in Nb particulate (also

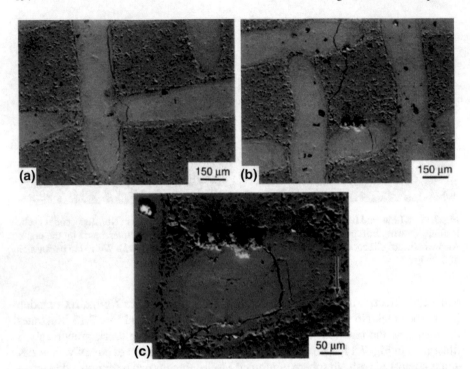

Fig. 7.13 Scanning electron micrographs of the crack/Nb wire interactions under cyclic fatigue loading in the $Nb_m/MoSi_2$ composite, showing **a** cracking along the weak reaction-layer interface and **b** and **c** failure of the Nb phase (which limits any ductile-phase bridging). Note the failue of the Nb ligament in **c** away from the primary crack plane and the presence of debris as a result of repetitive crack sliding at the interface. The arrow indicates the direction of crack growth. Badrinarayana et al. (1996). With kind permission of Springer. Nb_m stands for chopped wire mesh reinforcement

spherical particulate) strengthened $MoSi_2$ ($Nb_p/MoSi_2$) and clearly than in pure $MoSi_2$. Crack deflection and crack closure are the basic mechanisms responsible for the superior fatigue resistance. The influence of crack closure on the crack growth is illustrated in Fig. 7.18.

Figure 7.17 is represented in a more elucidating presentation together with the change of ΔK as a function of R in Fig. 7.19. Note that in 7.19b the ΔKth are in various cycles. Bridging was generally non-effective or even suppressed, because of the premature failure of the Nb wire. Crack growth rate increased also with increasing load ratio. Also growth rate was dependent on K_{max} and ΔK. As seen in Fig. 7.19b the ΔK versus R lines are quite close to each other in the R range 0.26–0.84. They deviate considerably at R values <0.26.

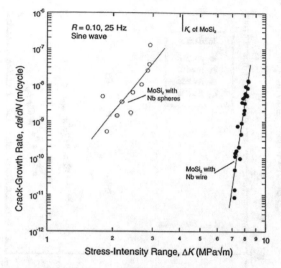

Fig. 7.14 Variation in cyclic fatigue-crack growth rates, da/dN, as a function of the nominal (applied) stress-intensity range, ΔK ($= K_{max} - K_{min}$), at R = 0.1 for $MoSi_2$ reinforced with 20 vol pct Nb in the form of high-aspect ratio with mesh ($Nb_m/MoSi_2$) and as spherical particulates ($Nb_p/MoSi_2$). Badrinarayana et al. (1996). With kind permission of Springer. Nb_m stands for chopped wire mesh reinforcement

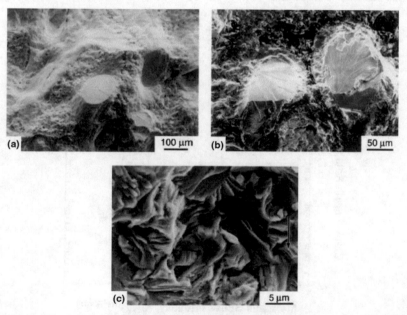

Fig. 7.15 **a** Scanning electron micrographs of the fatigue fracture surfaces in the $Nb_m/MoSi_2$ composite at R = 0.1, showing **a** morphology at low magnification and failure of a Nb wire **b** by transgranular cleavage and **c** by a quasi-cleavage mode. Note evidence of wear at the grain corners and debris on the fatigue surfaces. The arrow indicates the general direction of crack growth. Badrinarayana et al. (1996). With kind permission of Springer. Nb_m stands for chopped wire mesh reinforcement

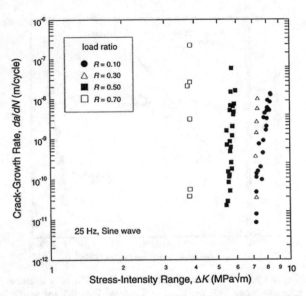

Fig. 7.16 Variation in cycle fatigue-crack growth rates, da/dN, for a range of load ratios from R = 0.1 to 0.7, in the $Nb_m/MoSi_2$ composite as a function of the applied stress-intensity range, $\Delta K = K_{max} - K_{min}$. Badrinarayana et al. (1996). With kind permission of Springer. Nb_m stands for chopped wire mesh reinforcement

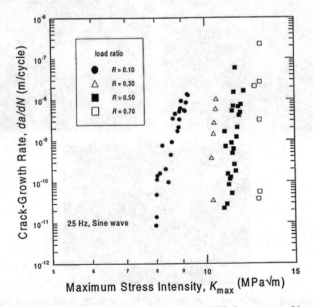

Fig. 7.17 Variation in cyclic fatigue-crack growth rates, da/dN, for a range of load ratios from R = 0.1 to 0.7, in the $Nb_m/MoSi_2$ composite as a function of the maximum stress intensity, K_{max}. Badrinarayana et al. (1996). With kind permission of Springer. Nb_m stands for chopped wire mesh reinforcement

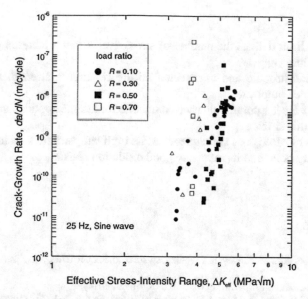

Fig. 7.18 Variation in cyclic fatigue-crack growth rates, da/dN, for a range of load ratios from R = 0.1 to 0.7, in the $Nb_m/MoSi_2$ composite as a function of the effective stress-intensity range, ΔK_{eff} (= $K_{max} - K_{cl}$), to allow for the influence of crack closure. Badrinarayana et al. (1996). With kind permission of Springer. Nb_m stands for chopped wire mesh reinforcement. K_{cl} stands for crack closure

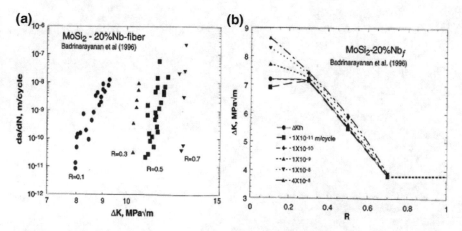

Fig. 7.19 Fatigue crack growth rates in $MoSi_2$-20% Nb_f: **a** crack growth rates in terms of ΔK; **b** ΔK variation with R for different crack growth. Sadananda et al. (1999). With kind permission of Elsevier

Summary

- Endurance limit defines the number of stress cycles that a material can undergo without failure
- To avoid catastrophic and unexpected failure requires a strengthened silicide, namely some composite
- $MoSi_2$-20% Nb is a commonly used composite for fatigue applications rather than the monolithic silicide
- The choice of $MoSi_2$ is a consequence of its high temperature structural stability, high melting point and its relatively good oxidation resistance.

References

K. Badrinarayana, A.L. McKelvey, K.T. Venkateswara Rao, R.O. Ritchie, Met. Trans. A **27A**, 3781 (1996)

J. Pelleg, in *Mechanical Properties of Ceramics* (Springer, 2014)

K. Sadananda, C.R. Feng, R. Mitra, S.C. Deevi, Mater. Sci. Eng. A **261**, 223 (1999)

W.O. Soboyejo, F. Ye, L.-C. Chen, N. Bahtishi, D.S. Schwartz, R.J. Lederich, Acta Mater. **44**, 2023 (1996)

K.T.Venkateswara Rao, W.O. Soboyejo, R.O. Ritchie, Met. Trans. A **23A**, 2249 (1992)

Chapter 8
Fracture in Silicides

Abstract Fracture in $CoSi_2$, $MoSi_2$, WSi_2 and $TiSi_2$ single and/or polycrystalline samples is considered in this chapter. Fracture is orientation and temperature dependent and occurs at some orientations even before the yield stress. An important design parameter is the fracture toughness determined by the stress intensity factor which can be evaluated from hardness measurements. Fracture toughness describes the ability of a material to resist fracture. These silicides are generally brittle but ductility sets in at some elevated temperature depending on the type of the silicide and its orientation. Catastrophic fracture should be eliminated by choosing the proper silicide, orientation and conditions of use.

8.1 Fracture in CoSi₂ Single Crystals

Catastrophic fracture at some orientations can occur in $CoSi_2$ single crystals even before the yield stress. Clearly the temperature level changes this tendency, and therefore its influence regarding plasticity before fracture is similar to that of the orientation. The single crystals prepared from rods of $CoSi_2$ were grown in an optical floating zone furnace in flowing atmosphere. Specimens for compression tests were cut from the as grown crystal with compression axes of the following orientations: [001], [011], [$\bar{1}11$], [$\bar{1}23$] and [$\bar{1}35$]. The stress strain curves at this orientations are shown in Fig. 8.1. A polycrystalline sample is also included in the figure. Note the effect of the orientation on the stress strain curves. At some orientations at room temperature fracture sets in immediately after a yield drop, in particular at orientations [011] and [$\bar{1}11$]. Note in particular that fracture sets in [001] without any sign of yielding similarly to the polycrystalline specimen and no evidence of slip lines were observed on surfaces of the fractured specimen. [$\bar{1}23$] and [$\bar{1}35$] specimens however indicated plastic deformation before fracture with a strain of 2–4%. On the fracture surfaces slip lines were observed. In the specimens showing the yield drop, namely [011], [$\bar{1}11$], a strain of ~1–2% before fracture was recorded. The primary slip system at room temperature is $\{100\}\langle100\rangle$ while the $\{001\}\langle\bar{1}10\rangle$ or the $\{111\}\langle\bar{1}10\rangle$ slip systems are not operative. Plastic deformation before fracture in the $\{100\}\langle100\rangle$ orientation is controlled by the Peierls' mechanism.

© Springer Nature Switzerland AG 2019
J. Pelleg, *Mechanical Properties of Silicon Based Compounds: Silicides*,
Engineering Materials, https://doi.org/10.1007/978-3-030-22598-8_8

Fig. 8.1 Stress-strain curves for CoSi$_2$ single crystals compressed at room temperature at a strain rate of 1×10^{-4} s^{-1}. Ito et al. (1992). With kind permission of Elsevier

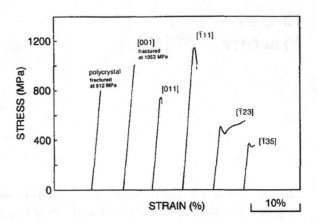

The temperature effect on the stress strain relation before fracture sets in is indicated for the orientations mentioned above seen in Fig. 8.2. As mentioned earlier [001] oriented single crystals fracture catastrophically without any sign of plastic deformation but at sufficiently high temperatures above 500 °C (in the Fig. 800 °C is indicated) plastic deformation occurs even in this orientation as seen in Fig. 8.2. The fracture strain as a function of temperature is illustrated in Fig. 8.3. As can be seen fracture strain rapidly increases with temperature increase.

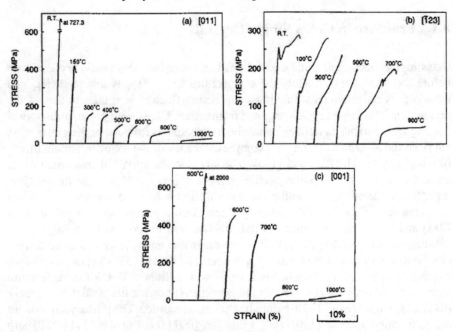

Fig. 8.2 Stress-strain curves of CoSi$_2$ single crystals with orientations **a** [011], **b** [$\bar{1}$23] and **c** [001] at a strain-rate of 1×10^{-4} s^{-1} at various temperatures. Ito et al. (1994). With kind permission of Elsevier

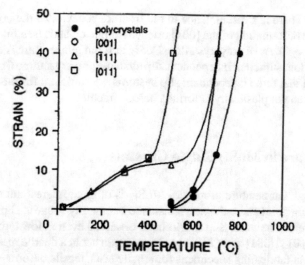

Fig. 8.3 Temperature dependence of compressive fracture strain for CoSi₂ single crystals with orientations [011], [$\bar{1}$11] and [001], and polycrystals. Ito et al. (1994). With kind permission of Elsevier

8.2   Fracture in Polycrystalline CoSi₂

Compression tests were performed on polycrystalline CoSi₂ also. Figure 8.4 shows the stress-strain relation in the temperature range RT-1000 °C. Similarly to the observations in single crystals at sufficiently high temperature, as a matter of fact above 500 °C plastic deformation sets in. The fracture strain increases with temperature as shown in Fig. 8.3 for polycrystals also. Above 500 °C polycrystalline CoSi₂ can be plastically deformed since in addition to the primary slip system on {001}⟨100⟩, slip systems of {111}⟨110⟩ and {110}⟨110⟩ become operative. This indicates hat for ductility in polycrystalline CoSi₂ the contribution of the secondary slip systems

Fig. 8.4 Stress-strain curves of CoSi₂ in polycrystals at a strain rate of 1×10^{-4} s^{-1} at various temperatures. Ito et al. (1994). With kind permission of Elsevier

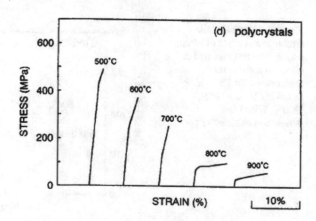

are essential. There is a resemblance in the fracture behavior of the polycrystalline CoSi$_2$ and the deformation of the [001] oriented single crystals (see Fig. 8.2c). Thus the fracture toughness of polycrystalline CoSi$_2$ at room temperature is quite low. As indicated without sufficient independent slip systems polycrystalline CoSi$_2$ (above it was indicated that three independent slip systems are sufficient for the introduction of ductility) can not plastically deformed before fracture.

8.3   Fracture in MoSi$_2$ Single Crystals

Due to the high temperature properties MoSi$_2$ is of great interest not only because its high melting point and oxidation resistance but mainly because slip deformation is expected before fracture. Slip in MoSi$_2$ is observed even at low temperatures on $\{011\}\langle100]$, $\{013\}\langle33\bar{1}]$ and $\{110\}\langle\bar{1}11]$ thus it can act in a ductile manner. Despite the difficulty in fabricating specimens for tensile tests, tensile deformation can shed light on the fracture process and provide understanding on ductility.

Single crystals of MoSi$_2$ were grown from ingots using an optical floating zone apparatus at a rate of 5 mm h^{-1} under argon flow. Figure 8.5 shows the tensile stress strain curves at various temperatures. Appreciable improvement in tensile elongation is seen in the figure at 1200 °C compared to the low temperature fracture without any ductility. The stereographic projection and the crystallographic orientation of the MoSi$_2$ single crystal are seen in Fig. 8.6. Note that the projections are for the C11$_b$ and C40 structures since they have similar atomic arrangement on (110) in C11$_b$ and (0001) in C40. The temperature dependence of elongation in tension and fracture strain in compression are seen in Fig. 8.7. As seen, in tension elongation increases at around 1100 °C from a zero brittle condition indicating the DBTT (ductile to brittle transition temperature).

In compression the fracture strain increases up to 600 °C, but then decreases, and again increases at around 1000 °C up to a plateau above this temperature and around 1200 °C. The operative slip system changes from $\{110)\langle\bar{1}11]$ to $\{011)\langle\bar{1}00]$. It might

Fig. 8.5 Typical tensile stress-strain curves of MoSi$_2$ single crystals deformed at different temperatures. Deformation at 1200 °C up to failure. Nakano et al. (2002). With kind permission of Elsevier

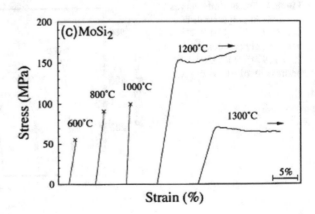

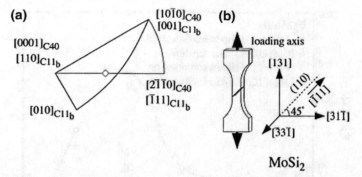

Fig. 8.6 Stereographic projections. **a** Showing a loading axis and specimen geometries and **b** showing crystallographic orientation for MoSi₂ single crystal. The projections for both C11ᵦ and C40 structures were overlapped due to the similar atomic arrangement on (110) in C11ᵦ and (0001) in C40 as shown in (**a**). Nakano et al. (2002). With kind permission of Elsevier. Note the C40 crystallographic projections of NbSi₂ having slip on {0001}<2Ī10> are not shown

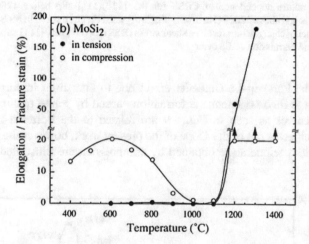

Fig. 8.7 Temperature dependence of elongation in tension and fracture strain in compression for MoSi₂ single crystal. Nakano et al. (2002). With kind permission of Elsevier

also be of interest to show the CRSS (critical resolved shear stress). It is shown in Fig. 8.8 for each slip system. The facture stress is indicated by crosses as seen in this figure.

MoSi₂ generally suffers from low fracture toughness at room temperature. However, MoSi₂ shows a strong plastic anisotropy, but in soft orientations it can be plastically deformed down to room temperature. The slip systems {011}⟨100], {010}⟨100] and {110}⟨111] exhibit a flow stress anomaly with a flow stress peak at ~800 °C for the first systems and at 1100 °C for the later one. In the range 700–1000 °C the {110}⟨111] system has the lowest critical flow stress. The flow stress anomaly was interpreted either by dislocation locking because of dissociation on a cross-slip

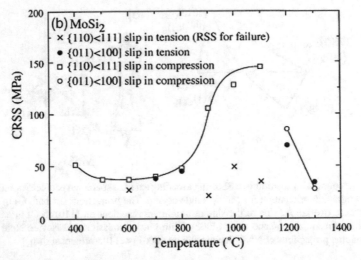

Fig. 8.8 Temperature dependence of CRSS for the $\{110\}\langle\bar{1}11]$ slip below 1200 °C and the $\{011\}\langle100]$ slip at and above 1200 °C in MoSi$_2$ single crystals. Fracture stress in MoSi$_2$ is described as a cross after being changed into resolves shear stress (RSS) for the $\{110\}\langle\bar{1}11]$ slip. Nakano et al. (2002). With kind permission of Elsevier

plane or by the Portevin-Le Chatelier effect due to interstitial impurities. (Recall that the effect is inhomogeneous deformation caused by solute pinning). Serrated stress-strain curves as seen in Fig. 8.9 are related to the Portevin-Le Chatelier effect. This subject is out of the scope of the present work, however the stress-strain curves at various temperature obtained by compression are illustrated for various

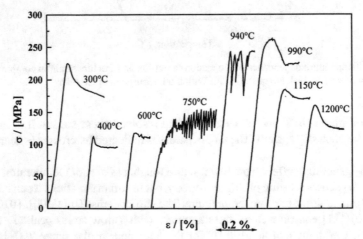

Fig. 8.9 Stress-strain curves of MoSi$_2$ with a [201] compression axis deformed at different temperatures at a strain rate of 10^{-5} s^{-1}. Guder et al. (1999). With kind permission of Elsevier

temperatures including serrated curves. The deformation was performed at a strain rate of 10^{-5} s^{-1}.

Compare Figs. 8.5 and 8.9, the former in tension and the later in compression. The anomaly manifests itself in the increase of the flow stress in the 600–1000 °C. At the lower and the higher temperature the stress decrease as usually observed due to the effect of increasing temperature.

8.4 Fracture in Polycrystalline MoSi$_2$

At low temperatures polycrystalline MoSi$_2$ is brittle similar to the characteristics of ceramics. At room temperature the fracture is 75% transgranular and 25% intergranular and the fracture toughness is 3 MPa m$^{1/2}$. These fracture modes are listed in Table 8.1 for various temperatures; grain size, fracture toughness and Vickers Hardness are included.

The experimental data were the result of Vickers hardness tests. The fracture toughness by indentation (hardness measurement) is calculated from the relation

$$K_c = A(E/H)^{1/2}(P/c^{3/2})$$ (8.1)

A is a constant with a value of 0.016, H is hardness, P the indentation load, E is the elastic modulus of MoSi$_2$ for polycrystalline 440 GPa and c is the average length of the four surface radial cracks formed during Vickers indentation fracture. The Vickers hardness is related to the load by

$$H_V = P/(2a^2)$$ (8.2)

where a is the half indent diagonal length. The Vickers indentation is illustrated in Fig. 8.10. Transgranular fracture inside of an individual grain is illustrated in Fig. 8.11. The crack is discontinuous. The cleavage in MoSi$_2$ is predominantly trans-

Table 8.1 MoSi$_2$ hot pressing temperatures and resulting grain size and fracture properties. Wade and Petrovic (1992a, b). With kind permission of John Wiley and Sons

Hot-press temperature (°C)	Grain size (µm)	Fracture toughness K_c (MPa m$^{1/2}$)	Vickers hardness, H_V (GPa)	Transgranular fracture (%)	Intergranular fracture (%)
1500	13.5	3.0	9.73	66.0	34.0
1600	15.3	3.6	9.87	70.9	29.1
1700	18.4	2.7	9.07	64.1	35.9
1800	20.5	3.0	9.08	67.3	32.7
1880	22.2	2.9	9.12	84.0	15.4
1920	31.9	2.3	8.92	97.0	3.0

Fig. 8.10 Typical Vickers indentation fracture pattern in MoSi$_2$ at 98-N (10-kgf) load under polarized light. Wade and Petrovic (1992a, b). With kind permission of John Wiley and Sons

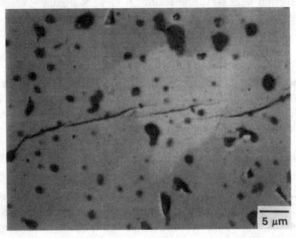

Fig. 8.11 Transgranular fracture morphology in MoSi$_2$ at 98-N (10-kgf) indentation load under polarized light. Wade and Petrovic (1992a, b). With kind permission of John Wiley and Sons

granular. Table 8.2 summarizes the fracture properties under different indentation loads. The predominant transgranular fracture obtained as illustrated in Table 8.1 means that the crystal cleavage planes are weaker than the grain boundaries, otherwise intergranular fracture would be the preferential route of the crack propagation. The suggested reason for the transgranular fracture is the crystallographic anisotropy and the layered structure in MoSi$_2$. The ratio c/a in the tetragonal MoSi$_2$ is large of 2.45 which is expected to generate anisotropic stresses in the individual grains after fabrication and cooling which contribute to the anisotropy. The estimated residual stress at room temperature in MoSi$_2$ having a grain size of 80 μm is high at a value of 84 MPa. The layered structure of MoSi$_2$, namely the layering of atoms may result in

Table 8.2 Fracture properties at different indentation loads for MoSi$_2$. Wade and Petrovic (1992a, b). With kind permission of John Wiley and Sons

Indentation load, P(N, (kgf))	c/a	Fracture toughness, K_c (MPa m$^{1/2}$)	Vickers hardness, H_V (GPa)
9.8 (1)	2.6	2.4[a]	9.4[a]
49 (5)	2.9	2.9	8.8
98 (10)	3.4	2.8	9.2
196 (20)	3.5	3.1	8.8
294 (30)	3.8	3.0	8.4
490 (50)	4.0	3.2	8.2
Average		3.0	8.7

[a]Value not included in average calculation

the existence of low energy cleavage planes parallel to the layers. These two aspects, namely the anisotropic internal residual stress from cooling during fabrication and the existence of low energy cleavage planes might explain the room temperature brittle transgranular fracture of MoSi$_2$. It might be of interest to illustrate an additional figure as Fig. 8.12, which outlines the grains in MoSi$_2$ and the crack propagating across the grain boundaries.

In situ straining experiments were performed in a high voltage transmission microscope (HVEM) using a special tensile stage for high temperatures. Plastic flow in the temperature range 495–1250 was observed and stress-strain curves are seen in Fig. 8.13 at the temperatures indicated. Up to 950 °C small work hardening is seen in the curves of Fig. 8.13, but above this temperature softening occurs as expected. Optical microscopy has shown slip steps below 1500 °C as a result of dislocation glide, but above this temperature slip traces were not observed. In Fig. 8.14 slip traces are shown. In the temperature range 900–100 °C slip traces appear in almost

Fig. 8.12 Close-up of radial crack fracture in MoSi$_2$ hot pressed at 1880 °C. Wade and Petrovic (1992a, b). With kind permission of John Wiley and Sons

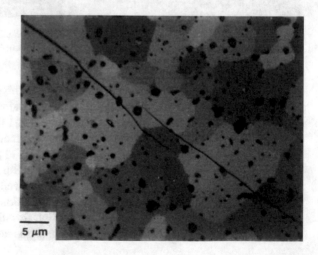

5 µm

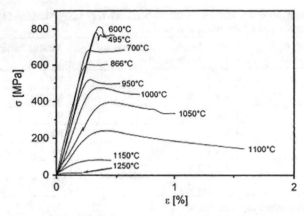

Fig. 8.13 Stress-strain curves at a strain rate of 2.5×10^{-7} s^{-1}. Junker et al. (2002). With kind permission of Elsevier

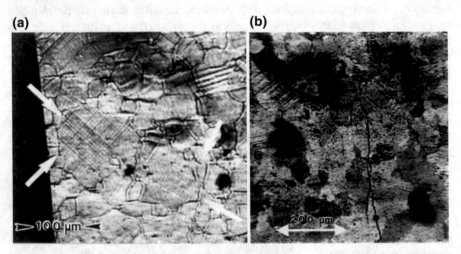

Fig. 8.14 Optical micrograph of deformed MoSi$_2$. Specimens deformed at 1000, 980 and 960 °C (**a**) and one deformed at 495 °C (**b**). Junker et al. (2002). With kind permission of Elsevier

all grains (Fig. 8.14a) and at lower temperature the number of slip traces decrease as seen in Fig. 8.14b. A crack is also seen crossing all grains in its wake. In Fig. 8.14a two sets of crossing slip steps are seen, but not in Fig. 8.14b. The traces represent slip on {011} and {110} planes independent of the deformation temperature. Analysis shows that most of the dislocations belong to the {011}⟨100⟩ slip system. {110} slip planes originate from dislocations with 1/2⟨111⟩ Burgers vectors. In Fig. 8.15 a micrograph of a specimen deformed at 1200 °C followed by brittle fracture is seen. Thus, below 1000 °C deformation occurs by dislocation glide on {110}⟨111⟩ and {011}⟨100⟩ slip systems. The von Mises criterion for slip is not satisfied by these two slip systems, the deformation is not homogeneous and microcracks are formed

Fig. 8.15 SEM micrograph
of a specimen deformed at
1200 °C followed by brittle
fracture at room temperature.
Junker et al. (2002). With
kind permission of Elsevier

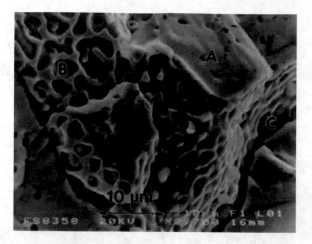

at the lowest temperatures as intergranular and transgranular cracks. Above 1000 °C
the deformation is controlled by visco-elastic grain boundary slide and decohesion.

8.5   Fracture in WSi$_2$ Single Crystals

Contrary to MoSi$_2$ where plastic flow is observed even at room temperature, brittle
fracture sets in single crystal WSi$_2$ in compression tests below ~1000 °C. The onset
temperature for plastic flow in WSi$_2$ depends on orientation. Thus, for example for
[$\bar{1}$10] it is 1100 °C and for [001] it is 1400 °C. A manifestation of the orientation
dependence of the flow stress can be seen in the stress strain curves shown in Fig. 8.16
for the orientations indicated. The range of temperatures onset for plastic flow is
clearly seen in the figure. The yield stress seen in Fig. 8.17 is at 0.2% plastic strain
in the stress-strain curves shown in Fig. 8.16. Note that x and + indicate stresses at
which fracture occurred in specimens of orientation [001] amd [$\bar{1}$10], respectively.
The lowest yield stress variation with temperature is obtain for orientation [$\bar{1}$10]. Note
the work softening at [001] orientation in Fig. 8.16d (similar to [0 15 1] orientation
in Fig. 8.16a) where the fracture strain is very limited to about 2% even at 1500 °C.
Fracture at this orientation always occurs catastrophically (Figs. 8.18 and 8.19).

8.6   Fracture in Polycrystalline TiSi$_2$

The crystal structure of TiSi$_2$ were illustrated in Figs. 2.8, 2.10 and 2.11. These figures
represent the base centered orthorhombic, C54 and C49 structures. Below the C54 and
C49 TiSi$_2$ structures are re-illustrated for convenience. Fracture toughness describes
the ability of a material to resist fracture, and is one of the most important properties of

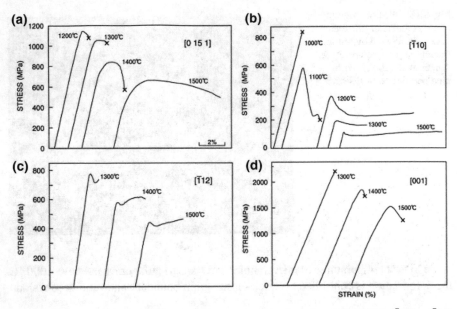

Fig. 8.16 Stress-strain curves of WSi$_2$ single crystals with orientations **a** [0 15 1], **b** [$\bar{1}$10], **c** [$\bar{1}$12] and **d** [001] at selected temperatures. Ito et al. (1999)

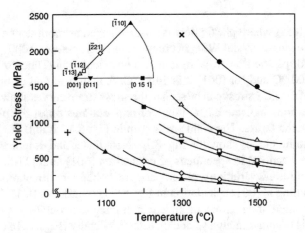

Fig. 8.17 Temperature dependence of yield stress for WSi$_2$ single crystals with orientations [001], [$\bar{1}$13], [$\bar{1}$12], [$\bar{2}$21], [$\bar{1}$10], [011] and [0 15 1]. The yield stress at which fracture occurred for [001] and [$\bar{1}$10] orientations are indicated by x and +, respectively. Ito et al. (1999)

a material for design applications. The fracture toughness of a material is determined from the stress intensity factor (K$_c$) at which a thin crack in the material begins to grow. K$_c$ denotes the crack opening mode. It is generally denoted as K$_{Ic}$ and is a measurement of the energy required to grow a thin crack. It is the stress x$\sqrt{\text{distance}}$ and its units are MPa $\sqrt{\text{m}}$. For the fracture toughness often an expression is given as:

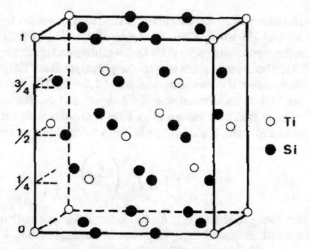

Fig. 8.18 TiSi₂ face centered orthorhombic C54 crystal structure. After Thomas et al. (1987)

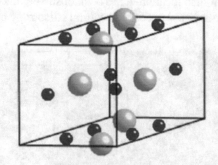

Fig. 8.19 TiSi₂ unit cell of C49. After Niranjan (2015)

$$K_{Ic} = \sigma \sqrt{\pi a B} \qquad (8.3)$$

Here, a is the crack length and B is a dimensionless parameter. From Eq. (8.3), the critical stress, σ_f, is:

$$\sigma_f = \frac{K_{Ic}}{\sqrt{\pi a B}} \qquad (8.4)$$

In TiSi₂—as in most materials—the fracture toughness and brittle microcracking are controlled by the density of local microvoids, the residual stress and the anisotropy of the linear thermal expansion coefficients. The microcracks are the origin of the transgranular fracture in materials.

TiSi$_2$ consolidation was by pulsed current activated combustion from Ti and Si powders achieving a high density TiSi$_2$. SEM micrographs in Fig. 8.20 illustrate the Ti–Si system at three processing stages. Vickers hardness and crack propagation are seen in Fig. 8.21. The average grain size of the polycrystalline TiSi$_2$ obtained by the pulsed current activated combustion was about 1.2 μm. The average hardness measured by the Vickers indentation was 964 kg/mm^2 and the fracture toughness 2.9 MPa m$^{1/2}$. Theses data were evaluated by Kim et al. who analyzed the fracture toughness according to Anstis et al. (1981) relating to the hardness according to (8.5).

$$K_{Ic} = 0.016 \left(\frac{E}{H} \right)^{1/2} \left(\frac{P}{c_0^{3/2}} \right) \tag{8.5}$$

Here E is the Young modulus, H is the hardness, P the indentation load and c_0 the crack trace length measured from the center of the indentation. It can be seen from relation (8.5) that the elastic modulus influences the fracture toughness. Further, grain boundaries play an important role in the crack propagation not only because often crack formation and propagation is initiated at boundaries but also because impurity segregation in such location. It has been observed that crack initiation takes place by intrinsic stresses based not only structural anisotropy but also in addition on

Fig. 8.20 Scanning electron microscope images of Ti + Si system: **a** after milling, **b** before combustion synthesis, **c** after combustion synthesis. Kim et al. (2009). Allowed as indicated by document

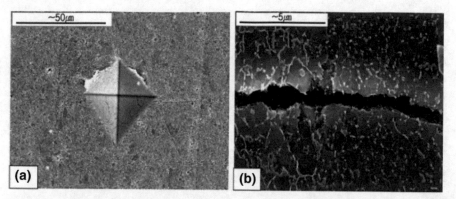

Fig. 8.21 **a** Vickers hardness indentation and **b** median crack propagating in TiSi$_2$. Kim et al. (2009). Allowed as indicated by document

impurities if present. Grain boundaries are the locations of various impurities among them oxides and segregated impurities decrease K$_{Ic}$. However the effect of the grain size by itself (when no impurities) can be observed by the increase of toughness by using ultrafine grained powders for the processing. Silicides are no exemption.

The temperature dependence of the fracture toughness was obtained from four point bend tests measured in the temperature range 25–1000 °C in air. The effect of temperature on the K$_{Ic}$ values is shown in Fig. 8.22. Note that the fracture toughness variation with temperature of Ti$_5$Si$_3$ is also included in Fig. 8.22. As can be seen the fracture toughness is low in the room temperature ~800 °C range (except to a small increase in the temperatures between them) but then a large increase in the fracture toughness occurs above the brittle to ductile transition temperature of TiSi$_2$. K$_{Ic}$ of TiSi$_2$ at room temperature is 1.9 MPa m$^{1/2}$.

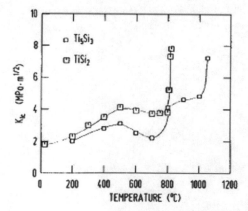

Fig. 8.22 Stress intensity factors of the intermetallic Ti$_5$Si$_3$ and TiSi$_2$ compounds as a function of test temperature. Rosenkranz and Frommeyer (1992). With kind permission of Elsevier

Summary

- Fracture is orientation and temperature dependent
- Only at certain orientations plastic flow occurs in silicides before fracture
- Fracture toughness is an indication for resistance to fracture
- Ductility before fracture is induced at elevated temperature deformation
- Transgranular fracture characterizes polycrystalline silicides.

References

G.R. Anstis, P. Chantikul, B.R. Lawn, D.B. Marshall, **64**, 533 (1981)

S. Guder, M. Bartsch, M. Yamaguchi, U. Messerschmidt, Mater. Sci. Eng. A **261**, 139 (1999)

K. Ito, H. Inui, T. Hirano, M. Yamaguchi, Mater. Sci. Eng. A **152**, 153 (1992)

K. Ito, H. Inui, T. Hirano, M. Yamaguchi, Acta Metall. Mater. **42**, 1261 (1994)

K. Ito, T. Yano, T. Nakamoto, H. Inui, M. Yamaguchi, Acta Mater. **47**, 937 (1999)

L. Junker, M. Bartsch, U. Messerschmidt, Mater. Sci. Eng. A **238**, 181 (2002)

B.-R. Kim, K.-S. Nam, J.-K. Yoon, J.-M. Doh, K.-T. Lee and I.-J. Shon, J. Cer. Proc. Res. **10**, 171 (2009)

T. Nakano, M. Azuma, Y. Umakoshi, Acta Mater. **50**, 3731 (2002)

M.K. Niranjan, Mater. Res. Express **2**, 1 (2015)

R. Rosenkranz, G. Frommeyer, Mater. Sci. Eng. A **152**, 288 (1992)

O. Thomas, R. Madar, J.P. Senateur, J. Less-Common Met. **136**, 175 (1987)

R.K. Wade, J.J. Petrovic, J. Amer. Ceram. Soc. **75**, 3760 (1992a)

R.K. Wade, J.J. Petrovic, J. Amer. Ceram. Soc. **75**, 1682 (1992b)

Chapter 9
Deformation in Nano Silicides

Abstract Strength properties are size dependent and increase with decreasing dimensions. The mechanical properties sharply deviate from those of macro-scale structures. It is expected to obtain improved strength properties in nano structures. Because the dislocation motion—if present at all—is restricted theoretical strength is anticipated. Contrary to macrostructures increase in strength is accompanied by increase in elongation. The theoretical tensile strength of $MoSi_2$ and WSi_2 were calculated and hardness measurements were measures. The hardness of β-$FeSi_2$ which is a candidate to be applied as light emitting diode were measured by Berkovich indenter.

9.1 Introduction

A nano particle—a three dimensional entity—on nanoscale is between 0.1–100 nm. The term of nano particle describes a small sized material, the mechanical properties of which sharply deviate from those of macro-scale structures. Therefore, knowledge of nano materials is important for the understanding of their mechanical properties, which dictate their practical applications. The strength properties of a structure is size dependent and they increase when its dimensions are small. Thus it is expected to obtain improved strength properties in nano structures. It is believed that dislocation motion—if present at all—is restricted. In such case when no dislocation motion is possible a material approaches its theoretical strength and thus to induce strain by deformation high stress is required. Contrary to macroscopic specimens where stress increase is associated with low ductility, in nanostructures a combination of high strength and good elongation is the general deformation pattern. The mechanical properties discussed in earlier chapters for macroscale silicides, will be reviewed here for nano seized silicides. We shall attempt—pending on available research data—to consider static, time dependent and cyclic properties, and fracture in nano structures. But before doing so it is of interest to show an often quoted schematic illustration of a one-dimensional nanostructure depicted by Gleiter and computed by Morse potential. The crystallite atoms in Fig. 9.1 are the open circles and the boundary regions are presented by dark circles. The atomic structure of all crystallites is identical with

© Springer Nature Switzerland AG 2019
J. Pelleg, *Mechanical Properties of Silicon Based Compounds: Silicides*,
Engineering Materials, https://doi.org/10.1007/978-3-030-22598-8_9

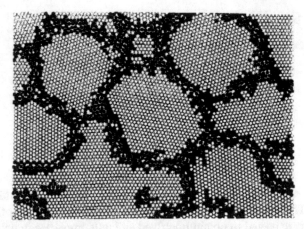

Fig. 9.1 Computed atomic structure of a nanostructured material. The computations were performed by modeling the interatomic forces by a Morse potential. The black (boundary) atoms are sites of which deviate more than 10% from the corresponding lattice sites. Gleiter (1995). With kind permission of Elsevier

the only difference between them being their crystallographic orientation. In the boundary regions, however the average atomic density and the coordination between nearest neighbor atoms deviate from the ones in the crystallites.

9.2   Tension

No data on tensile stress on the silicides considered in this book are available in nanostructures. It was felt that theoretical tensile strength of some transition metal disilicides with $C11_b$ structure, specifically $MoSi_2$ and WSi_2 ideal crystals, might provide data with probable equivalence to those to be expected in nanocrystals also. Theoretically, the ideal strength can be investigated using ab initio electronic structure methods based on the density functional theory. During the simulation complete relaxation of both external and internal parameters are allowed. The theoretical tensile strength of these silicides are determined for the [001] loading axis. The results are shown in Table 9.1.

Table 9.1 Theoretical tensile strengths σ_{th} of $MoSi_2$ and WSi_2. E_{001} is the Young modulus for the [001] loading. From Table III of Friák et al. (2003). With kind permission of Professor M. Friák

Material	Structure	σ_{th} (GPa)	σ_{th}/E_{001}	Reference
$MoSi_2$	$C11_b$	37	0.078	This work
WSi_2	$C11_b$	38	0.079	This work

9.3 Hardness Tests

9.3.1 Hardness in MoSi₂

Micro/nano hardness measurement were performed on as received and deformed monolithic $MoSi_2$. The purpose of the deformation was to see if prestaining influences the hardness values. The microstructure of the $MoSi_2$ is illustrated in Fig. 9.2.

Illustrations of load versus depth of indentation are presented in Figs. 9.3 and 9.4 for as as received specimen and with 50 and 500 mN loads. In Figs. 9.5 and 9.6 the as received and deformed specimens under 500 and 50 nM loads are compared. Surprisingly, one observes from the data of Figs. 9.5 and 9.6 that the deformed specimens are softer than the as received ones, since the penetration depths are larger. Further a definite load effect is observed as seen from the illustrations. The microhardnesses of the as-received and as-deformed specimens under 50 and 500 mN loads are listed in Table 9.2. HU in the table refers to the so-called universal hardness defined as

$$HU = F/28.43\,h^2 \qquad (9.1)$$

where F is the test force divided by the apparent area of indentation A(h) under the applied test force. The conventional Vickers hardness can be expressed as (Pelleg 2014)

$$HV = 1.8544\,F/a^2 \qquad (9.2)$$

where a is the diagonal of the indentation in mm. Values of the Vickers hardness are presented in Table 9.3.

One should note that in Figs. 9.3 and 9.4 values of HU and HU_{plast} are illustrated. HU_{plast} refers to the plastic hardness which is the quotient of the maximum test force, F_{max}, divided by an area calculated by the extrapolation and is given as

Fig. 9.2 Microstructure of $MoSi_2$. Henzel et al. (2004). With kind permission of Springer

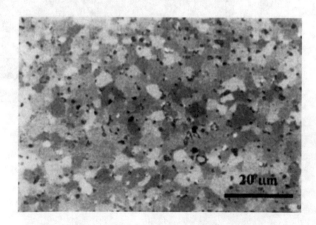

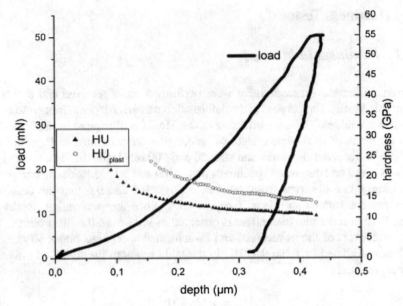

Fig. 9.3 Depth-sensing curve of the as-received state with the maximum load of 50 mN. Henzel et al. (2004). With kind permission of Springer

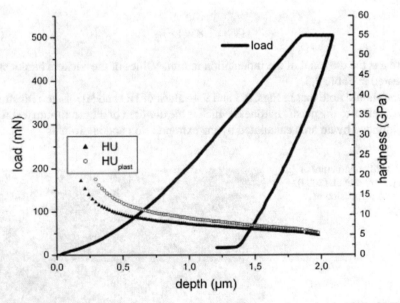

Fig. 9.4 Depth-sensing curve of the as-deformed state with the maximum load of 500 mN. Henzel et al. (2004). With kind permission of Springer

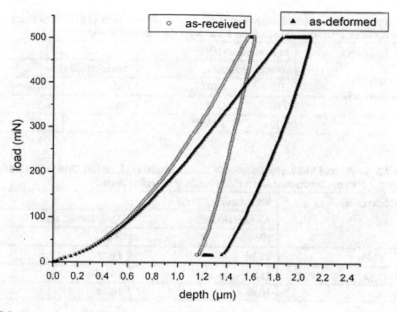

Fig. 9.5 Comparison of the P-h curves for the as-received and as-deformed states with the maximum load of 500 nM. Henzel et al. (2004). With kind permission of Springer

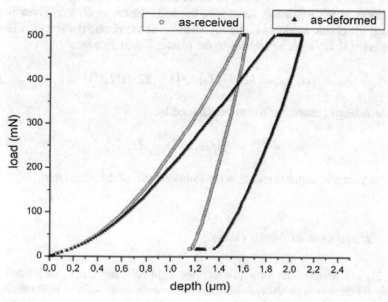

Fig. 9.6 Comparison of the P-h curves of the as received and as-deformed states with the maximum load of 50 nM. Henzel et al. (2004). With kind permission of Springer

Table 9.2 Hardness values calculated from the P-h curves. Henzel et al. (2004). With kind permission of Springer. P stands for the load and h stands for the penetration depth

Max. load (mN)	Microhardness (GPa)			
	As-received state		As-deformed state	
	HU	HU$_{plast}$	HU	HU$_{plast}$
500	7.66	9.19	5.51	6.25
50	1.78	14.17	9.35	11.77

Table 9.3 Traditional Vickers microhardness values. Henzel et al. (2004). With kind permission of Springer. P stands for the load and h stands for the penetration depth

Max. load (mN)	Microhardness (GPa)	
	As-received state	As-deformed state
	HV	HV
500	13.34	11.78
1000	11.27	10.04
2000	10.84	10.44

$$HU_{plast} = F_{max}/26.43\, h_r^2 \tag{9.3}$$

h_r is the indentation depth resulting as the intersection of the tangent of indentation depth curve at the maximum force with the indentation depth axis given in mm. Knowing the elastic modulus E, the contribution of the elastic deformation HU can be converted to HV which is related to the plastic indent size by

$$HU_{plast} = 4HU/\left\{1 + \sqrt{(1 - 12HU/E^*)}\right\}^2 \tag{9.4}$$

E^* is the effective contact stiffness determined by

$$E^* = \left\{(1 - v_S^2)/E_S + (1 - v_i^2)/E_i\right\}^{-1} \tag{9.5}$$

E_S is the Young's modulus and v_S is the Poisson ratio of the tested material.

9.3.2   Hardness in Nano FeSi$_2$

FeSi$_2$ (β) is a semiconductor applicable for light emitting diode, which has been recently fabricated as nano crystalline material down up to 3–5 nm in diameter. The film was grown by the so-called facing target (FT) dc sputtering method. (The FT technique uses two parallel targets and the substrate is placed perpendicularly to the two targets outside the plasma formed between them). The interest in β-FeSi$_2$ is due to its being a promising material for optoelectronic and mechanical applications.

Berkovich indenter was used for the nano-indentation tests. Depth-sensing indentation has been made for evaluating the hardness of FeSi$_2$ thin films having a thickness of 500 nm coated on Si and glass substrates prepared at 0.5 Pa and 0.6 kW for 125 min. A typical indentation load displacement is shown in Fig. 9.7. The figure includes also

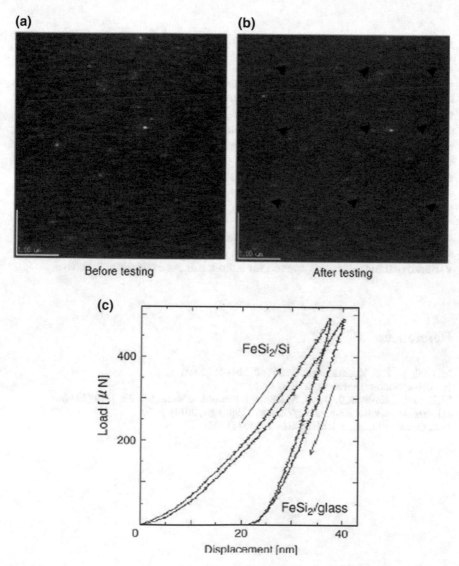

Fig. 9.7 AFM images of 5 × 5 μm^2 square before and after nano-indentation tests. Traces of Berkovich indenter are shown in (**a**). Load-displacement curves for FeSi$_x$ on both Si and glass substrates are shown in (**b**)

indentations on Si and glass. The depth was 1/10 of the film thickness so the effect of the substrate could be ignored. AFM images of 5×5 μm^2 before and after the indentations can also be seen in the figure. The resulting nano-indentation hardness is ~10 GPa. Vickers hardness can be evaluated from the load-displacement curves as indicated above in Eqs. 9.1–9.5, but in the case of the nano indentations of $FeSi_2$ the method given in the paper of Oliver and Pharr (1992) was used for the calculation.

9.3.3   Hardness in Nano $TiSi_2$

Despite of the distinct properties of $TiSi_2$ of low density high corrosion resistance, its application in VLSI, high melting point, hardness and creep strength no nano-structure data are available in the literature. Its high hardness of ~850 HV and its flow stress more than 1200 MPa makes this material an important aero-space engineering material and one might expect an active research investigation of the nano structure which most likely would further enhance the mentioned properties mainly those of the mechanical nature such as hardness, stress etc.

Summary

- Calculated theoretical strength of $MoSi_2$ and WSi_2 is presented
- Micro/nano hardness indentations were performed on $MoSi_2$,
- Berkovich indenter were applied for nano-hardness evaluation of $FeSi_2$.

References

M. Friák, M. Šob, V. Vitek, Phys. Rev. B **68**, 184101 (2003)
H. Gleiter, Nanostructured mater. **6**, 3 (1995)
M. Henzel, J. Kovalcik, J. Dusza, A. Juhasz, J. Lendvai, J. Mater. Sci. **39**, 3769 (2004)
J. Pelleg, *Mechanical Properties of Ceramics* (Springer, 2014), p. 79
W.C. Oliver, G.M. Pharr, J. Mater. Res. **7**, 1564 (1992)

Chapter 10
The Effect of B

Abstract Small additions of B improve the room temperature ductility of many intermetallic compounds $CoSi_2$, $MoSi_2$ and $TiSi_2$ among them. Segregation of B to grain boundaries strengthen them. The outstanding feature of B is on inducing ductility in brittle material. Also the oxidation resistance is improved as exemplified for $MoSi_2$ and pest formation is eliminated. Hardness and fracture toughness are higher in B added silicides as illustrated for $MoSi_2$. B added $TiSi_2$ coatings are an example of the effect of B additions, but no direct evidence of the effect of B on the mechanical properties exists, probably because efforts were directed to evaluate its use in VLSI.

10.1 Introduction

Large amount of experimental and theoretical work on the effect of B addition to material—in particular to intermetallic compounds—has been done, but its effect still remains controversial since no unified theory exists. Some of them—without entering into the specific details—propose: (a) the force between the atomic bonds change; (b) the nature of the atomic bonds change, namely from metallic to covalent bonds and (c) change in the metallic bonds in the immediate vicinity of solute atoms. These approaches attempt to explain the cohesion changes in the grain boundaries due to B (boron) segregation to grain boundaries which is observed experimentally when the B content exceeds a certain solubility value, the solubility limit.

Small additions of boron have been shown to improve the room temperature ductility of many intermetallic compounds. The boron has a strong tendency to segregate to grain boundaries and strengthen them, allowing the inherent ductility of the grains to be achieved. Further, it is believed that additions of B to some silicides is refining crystal grains and improving room-temperature fracture toughness. The improvement must reflects itself in the microstructure. The meaning of ductility increase by B addition is that the fabricability of alloys—usually brittle or difficult for forming—is much improved. The strong tendency of B to segregate to grain boundaries does not necessarily mean that the cohesive strength of the grain boundaries is enhanced, although some existing theories on solute segregation effects attribute the beneficial effect of B to grain boundary cohesion.

© Springer Nature Switzerland AG 2019 163
J. Pelleg, *Mechanical Properties of Silicon Based Compounds: Silicides*,
Engineering Materials, https://doi.org/10.1007/978-3-030-22598-8_10

As indicated no uniform theory exists regarding the mechanism of the B effect. One cannot disregard the concept by some investigators that the segregation of B is a factor affecting the mobility of dislocations and slip accommodation at grain boundaries.

10.2   B Effect in CoSi$_2$

The attention that CoSi$_2$ has received is mainly because its use in microelectronics as contact and interconnection. Also its attractive properties such as high strength and heat resistance are of interest in the application of CoSi$_2$. However polycrystalline CoSi$_2$ shows a high ductile brittle transition temperature. CoSi$_2$ is brittle and exhibits intergranular fracture as illustrated in Fig. 10.1. B doped CoSi$_2$ was obtained by melting together CoSi$_2$ and CoSi$_2$ containing B. The final B concentration was 0.038 at.% B. This level of B is within the range of its solubility level in the temperature range of 750–900 °C. Table 10.1 lists the chemical analysis of the fractured area obtained by AES (Auger electron spectroscopy). The uncertainty in composition is in the 0.01 at.% range. The interest in B addition is because B addition improves the mechanical properties specifically the ductility by decreasing grain boundary brittleness. At the level of B added the intergranular brittleness of CoSi$_2$—which is an intrinsic property—did not change drastically with B addition. B was observed to segregate to grain boundaries as indicated by AES in Fig. 10.2. The B small peak is seen in the AES at the peak of 179 eV was observed only in sample 4 listed in Table 10.1. The chemical grain boundary composition is comparable to the bulk composition and is not affected by the segregation of the added B.

Fig. 10.1 SEM micrograph of fracture surfaces of pure CoSi$_2$. Intergranular fractures and cracks are present ($\times$230). Malchère et al. (1991). With kind permission of Elsevier

Table 10.1 Chemical analysis of the fractured area by AES. Malchère et al. (1991). With kind permission of Elsevier

Sample		$C_{Si}^{surf(GB)}$ (at.%)	$C_{Co}^{surf(GB)}$ (at.%)	$C_{B}^{surf(GB)}$ (at.%)
1	CoSi$_2$	68	32	–
2	CoSi$_{2+\delta}$	72	28	–
3	CoSi$_{2-\delta}$	Contaminated fissure before fracture in AES		
4	CoSi$_2$ + B	64	34	2 ± 0.5
5	CoSi$_{2+\delta}$ + B	66.5	32	1.5 ± 0.5
6	CoSi$_{2-\delta}$ + B	67	31	2 ± 0.5

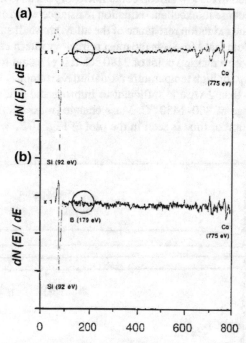

Fig. 10.2 Auger electron spectra from the fracture surface of **a** CoSi$_2$ and **b** boron-doped CoSi$_2$. Malchère et al. (1991). With kind permission of Elsevier

The amount of B segregation at the grain boundaries is about the same in all samples. The grain boundary segregation energy for B was calculated resulting in $\Delta G_B^{GB} = 55 \pm$ kJ mol^{-1} ($\cong -0.6$ eV/atom). ΔG_B^{GB} represents the decrease in energy when the B atom exchanges position from the bulk lattice to the grain boundary.

No mechanical change due to B addition is indicated, because the aim was to show that B segregates to grain boundaries.

10.3   B Effect in MoSi₂

During a sintering process, it is necessary to prohibit the occurrence of abnormal or elongated grain growth in order to obtain a higher density. The use of additives has been proven to be advantageous in this respect. For example, boron addition is one of additives considered for this purpose. But in addition it is expected to improve oxidation resistance of $MoSi_2$ and affect the strength and ductility properties. Due to its very high melting point, low density and outstanding oxidation resistant even without additives, the most important use to date of $MoSi_2$ is as heating elements known as Kanthal. Unfortunately the very low ductility and fracture toughness of the pure unalloyed compound has limited use for non-structural applications. Incorporation of small quantities of B ~ 0.8–1.9 wt% could considerably improve ductility, however the interest has shifted to the low silicon content Mo_5Si_3 because its high creep resistance and mainly because its excellent oxidation resistance when alloyed with boron. To get a feeling on the oxidation resistance of the alloy, we shall start with this observation in Mo_5Si_3. Note that Mo_5Si_3 is the most refractory intermetallic compound in the Mo-Si system with a melting point of 2180 °C. It is resistant to high temperature deformation and its poor high temperature oxidation resistance is improved by small addition of B. Less than 2 wt% is sufficient to improve the oxidation resistance in the temperature range of 800–1450 °C. Mass change (mass loss) occurring during oxidation as a function of time is seen in the plot of Fig. 10.3. Without B addition,

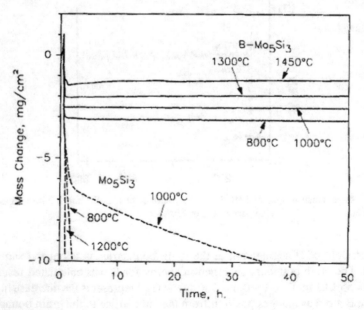

Fig. 10.3 Comparison of the oxidation of Mo_5Si_3 with B-Mo_5Si_3. B-Mo_5Si_3 traces are offset for clarity. Composition (wt%) of B-Mo_5Si_3 is 16.1 Si, 1.24 B, balance Mo. Meyer et al. (1996). With kind permission of John Wiley and Sons

pest behavior is observed at 800 °C, but no pest occurs at 1000 °C. However the oxidation is very rapid and increases at >1000 °C. Above this temperature it converts after a short time to SiO_2.

B addition forms a protective borosilicate layer, closes pores after an initial MoO_3 volatilization. No pesting occurs at 800 °C. Above 1000 °C oxidation is limited by oxygen diffusion through the scale. As the scale grows thicker, the diffusion distance for oxygen diffusion becomes longer, and the oxidation rate slows parabolically. At 1200 °C the oxidation rated decrease 10^5 times over Mo_5Si_3 (without B). The steady state mass change is shown in Fig. 10.4. As seen in this figure the oxidation rate is comparable to that of $MoSi_2$. The oxidation behavior as affected by the B concentration is illustrated in Fig. 10.5 for various M-Si-B compositions. As can be seen in the plots that after an initial mass loss due to molybdenum oxide volatilization the oxidation slows down markedly. The initial molybdenum oxide loss is larger for the Mo-rich composition. Figure 10.6 compares creep rates of monolithic $MoSi_2$ with that of sintered $B-Mo_5Si_3$. The creep rate was evaluated in the temperature range 1220–1320 °C and at loads of 140–180 MPa. The composition in wt% is 85.7 Mo, 13.0 Si, 1.3 B. The material is a Mo_5Si_3 matrix with Mo_3Si and $Mo_5(Si,B)_3$ as second phases.

The $Mo_5(Si,B)_3$ phase is known as T2. An average energy for creep was evaluated as 396 kJ/mol and the stress exponents are in the range 4.8–5.0. The multi-phase $B-Mo_5Si_3$ has a lower creep rate than most $MoSi_2$ matrix composites. In conclusion this section on Mo_5Si_3 it was observed that B substantially improves oxidation resistance over the temperature range 800–1450 °C and a significant decrease in steady state oxidation rate occurs. In a Mo_5Si_3 compound improved creep resistance was observed as compared with $MoSi_2$. B addition improves the oxidation resistance of $MoSi_2$ also.

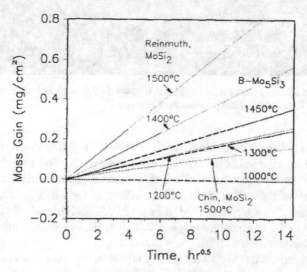

Fig. 10.4 Comparison of steady-state oxidation rates of $MoSi_2$ and $B-Mo_5Si_3$. Data is offset to intercept zero. Meyer et al. (1996). With kind permission of John Wiley and Sons

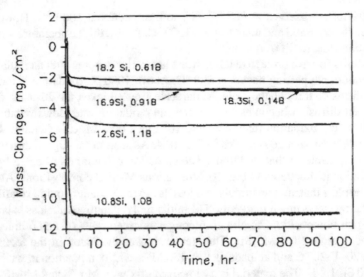

Fig. 10.5 Oxidation behavior as a function of composition for Mo-Si-B. Composition in wt%, balance is molybdenum. Meyer et al. (1996). With kind permission of John Wiley and Sons

Fig. 10.6 SEM micrographs of **a** Mo-67Si and **b** Mo-47Si-23B powders after milling for 20 h. Taleghani et al. (2014). With kind permission of Elsevier

Mo-Si in ratio 1:2 and B containing Mo-Si (30Mo-47Si-23B in atomic percent) were used to produce stoichiometric $MoSi_2$ and $MoB/MoSi_2$. The fabrication method of the alloys was by mechanical alloying with subsequent reactive sintering. The microstructure of these alloys after 20 h milling is seen in Fig. 10.6. The mass change in these alloys are compared in Fig. 10.7. Monolithic $MoSi_2$ is oxidized to a large extent in the temperature range of 550–800 °C, however in the B containing alloys pest disintegration of $MoSi_2$ was prevented.

The micrograph of the $MoB/MoSi_2$ alloy is seen in Fig. 10.8. Cross section micrographs of the oxidized alloys at 1000 °C is illustrated in Fig. 10.9. Fine particles of $MoSi_2$ are uniformly distributed in the MoB matrix of this $MoB/MoSi_2$ composite.

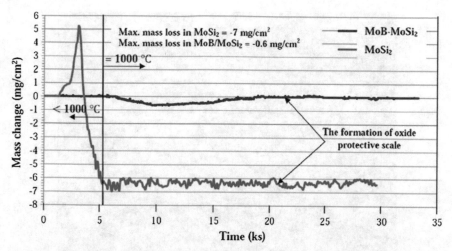

Fig. 10.7 Weight change versus time for the oxidation of MoB/MoSi₂ and monolithic MoSi₂ at 1000 °C in air atmosphere. Taleghani et al. (2014). With kind permission of Elsevier

Monolithic MoSi₂ was severely oxidized (Fig. 10.9b) with cracks in the temperature range 550–800 °C, B additions prevents pest integration in the MoB/MoSi₂ alloy because the preventive layer formed which retards oxygen diffusion during oxidation.

As mentioned B—one of the alloying elements—is added to MoSi₂ to improve properties. Alloys prepared by mechanical alloying with pulse discharge sintering (MA-PDS) showed high hardness and fracture stress due to their very fine grain sizes. The microstructure of the B added MoSi₂ is compared with the B free alloy in Fig. 10.10. The Vickers hardnesses of the MA-PBS alloy fabricated materials are compared for MoSi₂ and MoSi₂-B in Fig. 10.11. The hardness values of MoSi₂ with Al and Nb are also included in the figure. Further comparison is made between MA-PBS alloys with alloys fabricated in argon. Note that the 5 at.% B MoSi₂ alloy is harder than the pure MoSi₂. The fracture stress is depicted in Fig. 10.12 for both MoSi₂ produced from MA and commercial powders and are compared with B alloyed MoSi₂. The B containing MoSi₂ is much better than the unalloyed MoSi₂ showing a higher fracture stress. The fractographs of the tensile specimens are seen in Fig. 10.13. The sample made from commercial powder fractured in a transgranular mode. The monolithic MoSi₂ made by MA-PDS process showed also a transgranular fracture mode. The estimated grain size for the MA-PDS alloy is ~1 μm.

Apparently the B additive does not have to be in the elemental form to produce the beneficial effect on MoSi₂. It has been observed that boron oxide (B₂O₃) also modifies the MoSi₂ behavior. The effect can be seen in the densification of MoSi₂ which is a function of the B₂O₃ concentration as indicated in Fig. 10.14. The sintering of the MoSi₂ was at 1700 °C and the amount of the sintering additive B₂O₃ was in the range of 250–2000 ppm. The sintering time was 0.5, 1 and 2 h. The linear shrinkage for the sintering times indicated is seen in Fig. 10.15 as a function of the

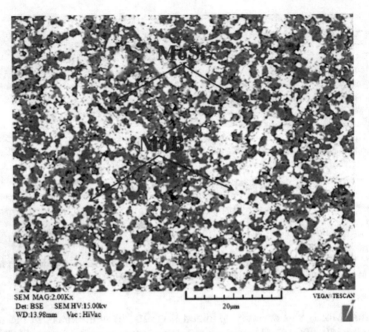

Fig. 10.8 The backscatter electron SEM micrograph from cross-section of as-mechanically alloyed M0-47Si-23B samples after sintering at 1300 °C for 3 h. Taleghani et al. (2014). With kind permission of Elsevier

B_2O_3 concentration. The maximum in Fig. 10.15 is at ~500 ppm B_2O_3. No abnormal or elongated grains was found when B_2O_3 was added, but agglomeration occurred during sintering without the additive as seen in Fig. 10.16. It is believed that part of B_2O_3 segregates at the grain boundaries forming a glassy phase and the rest remain in the bulk to hinder diffusion, thus decreasing the nominal diffusion coefficient. It is suggested that the additive alters surface energy to reduce the diffusion coefficient. The mechanism still remains controversial but the commonly accepted fact is that an additive retards abnormal or elongated grain growth and thus results in a fast pore removal rate to achieve high density.

10.4   B Effect in TiSi$_2$

B doped TiSi$_2$ coating is an excellent protection against isothermal and cyclic oxidation of commercially pure Ti and some of its alloys such as Ti-22Al-27Nb and Ti-20Al-22Nb which are high temperature refractory materials. Their use is limited because their poor oxidation resistance above 550 °C in air. These alloys undergo two kinds of deteriorations, (a) rapid consumption of the base metal to form an oxide scale and (b) embrittlement by the inward penetration of contaminants. Protective coating is one of the methods to protect Ti alloys by forming a slow growing scale.

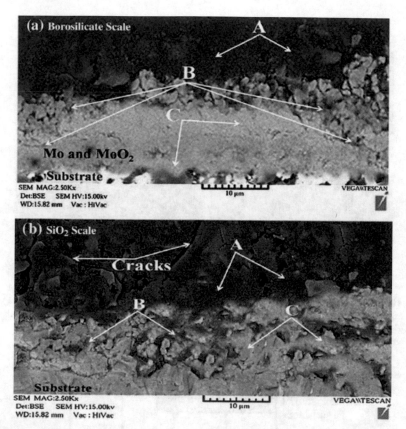

Fig. 10.9 Cross-sectional SEM micrograph of oxidized alloys at 1000 °C: **a** MoB/MoSi$_2$ composite and **b** monolithic MoSi$_2$. Taleghani et al. (2014). With kind permission of Elsevier

For the experimental method and details of the B containing TiSi$_2$ coating formation the work of Cockram and Rapp. should be read. The cross section of the B doped silicide coating on commercially pure Ti is seen in the optical micrograph of Fig. 10.17a, and its microprobe profile in Fig. 10.17b. A Si profile is also indicated in the microprobe profile since SiO$_2$ is a part of the protective scale. Micrographs of thick and thin protective coatings on one of the Ti alloys, namely the Ti-22Al-27Nb, is illustrated in Fig. 10.18.

The weight gain due to isothermal oxidation of the commercial Ti B coated TiSi$_2$ at 500–1000 °C is shown in Fig. 10.19. The significant weight gain occurring in the beginning of about 2 h oxidation, is followed by a slow parabolic growth rate kinetics. It might be of interest to show the Arrhenius plot for the parabolic kinetics oxidation of coating of B doped TiSi$_2$ in comparison with TiSi$_2$ and pure Si (Fig. 10.20). The weight gain in cyclic oxidation is seen in Fig. 10.21 at two temperatures, those of 800 and 1000 °C, compared with Ge-doped TiSi$_2$. Further, the same coatings at the same temperatures are seen also on Ti-22Al-27Nb substrates in the plots of Fig. 10.22 compared with the uncoated substrate Ti alloy.

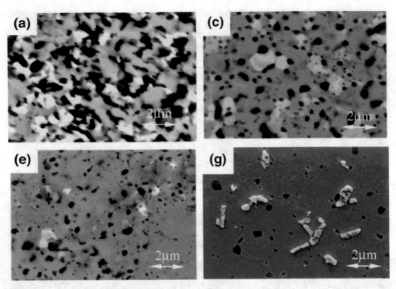

Fig. 10.10 SEM micrographs of MoSi$_2$ and MoSi$_2$-B in air: (**a**) and (**c**) and in argon (**e**) and (**g**) respectively. Shan et al. (2002). With kind permission of The Japan Institute of Metals and Materials

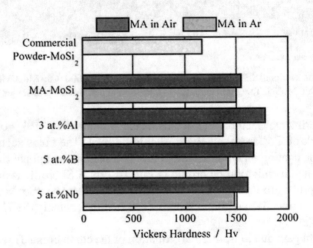

Fig. 10.11 Vickers hardness of MoSi$_2$ and MoSi$_2$-X (X = Al, B or Nb) alloy fabricated by MA-PDS process. The hardness of MoSi$_2$ sintered from commercial powders is also shown in the figure. Shan et al. (2002). With kind permission of The Japan Institute of Metals and Materials. MA stands for mechanical alloying

The oxygen and other contaminants penetration beneath the scale/metal interface can be detected in Fig. 10.23 by microhardness measurements and in Fig. 10.24 indicating a steep hardness gradient in the case of coating on Ti-22Al-27Nb. Neither

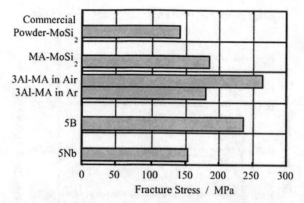

Fig. 10.12 Fracture stress of MoSi$_2$ and MoSi$_2$-X (X = Al, B or Nb) alloy fabricated by MA-PDS process. The fracture stress of MoSi$_2$ sintered from commercial powders is also shown in the figure. Shan et al. (2002). With kind permission of The Japan Institute of Metals and Materials. MA stands for mechanical alloying

Fig. 10.13 SEM micrographs of fracture surface tensile test of MoSi$_2$ and MoSi$_2$-B alloy fabricated by MA-PDS process. **a** MoSi$_2$ made from commercial powder milled in air, **b** and **c** are those of 5B and pure MoSi$_2$ milled in argon atmosphere. Shan et al. (2002). With kind permission of The Japan Institute of Metals and Materials

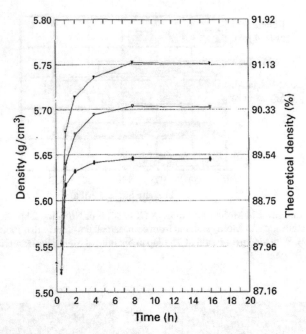

Fig. 10.14 Densification of MoSi$_2$ using various amounts of B$_2$O$_3$. Sintering temperature 1600 °C: ● 0 ppm; ▽ 250 ppm; ▼ 500 ppm. Ting (1995). With kind permission of Springer

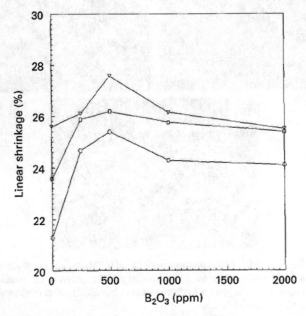

Fig. 10.15 Linear shrinkage of MoSi$_2$ versus the amount of B$_2$O$_3$ added during the sintering of MoSi$_2$: ○ 1/2 h samples; □ 1 h samples; ▽ 2 h samples. Ting (1995). With kind permission of Springer

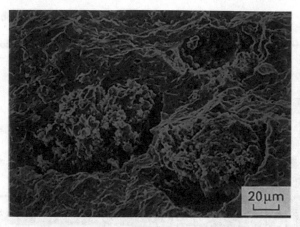

Fig. 10.16 Fractured surface of a specimen sintered at 1700 °C for 2 h with no B$_2$O$_3$ added. Ting (1995). With kind permission of Springer

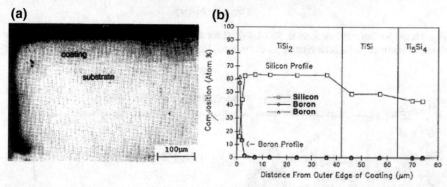

Fig. 10.17 Polished cross section of boron-doped silicide coating on commercial pure Ti in the coated condition: **a** optical micrograph, **b** microprobe profile of B and Si. Cockeram and Rapp (1995). With kind permission of Elsevier

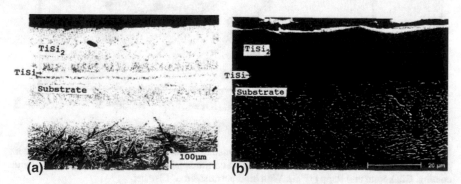

Fig. 10.18 Micrograph of the polished and etched cross-section of the boron-doped silicide coating: **a** optical micrograph of a thick coating on Ti-22Al-27Nb alloy; **b** SEM image of a thin coating on Ti-20Al-22Nb. Cockeram and Rapp (1995). With kind permission of Elsevier

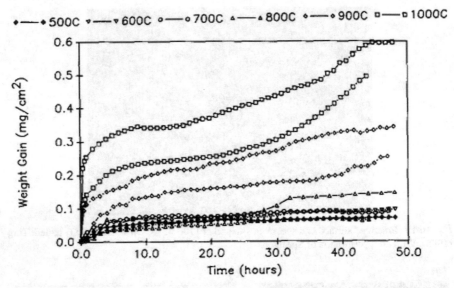

Fig. 10.19 Isothermal oxidation at 500–1000 °C for 48 h in air of coatings on commercial pure titanium boron-doped silicide coating. Cockeram and Rapp (1995). With kind permission of Elsevier

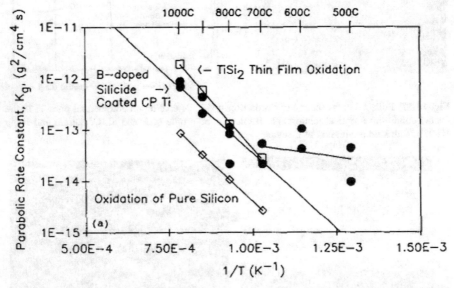

Fig. 10.20 Arrhenius plot of the parabolic rate constant versus inverse temperature compared with oxidation data of pure Si and TiSi$_2$ for coating on commercial pure titanium: boron doped silicide coating. Cockeram and Rapp (1995). With kind permission of Elsevier

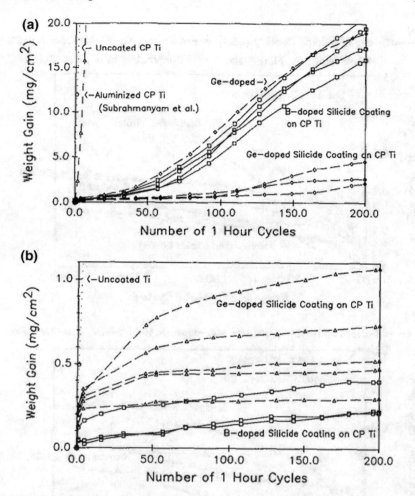

Fig. 10.21 Cyclic oxidation of boron-doped and germanium-doped silicide coating on commercial pure titanium and uncoated titanium for 200 cycles in air: **a** at 1000 °C, compared with data obtained from literature for a Ti_3Al diffusion coating on commercial pure titanium; **b** at 800 °C. Cockeram and Rapp (1995). With kind permission of Elsevier

hardness gradient nor change in the baseline hardness is observed following 200 oxidation cycles at 800 °C as seen in Fig. 10.23. Thus the B-doped $TiSi_2$ coating is a good and effective barrier against oxidation and inward penetration of oxygen or other contaminants. The sharp hardness gradient observed in the plots of Fig. 10.24 is a result of Al diffusion from the substrate and its penetration to the interface. Note that the hardness at zero distance from the coating/substrate interface might to some extent be representative of the B-doped $TiSi_2$ hardness at a value of ~100 kg/mm² Knoop hardness. The correctness of this conjecture has to be proven experimentally.

No direct data on the mechanical properties on B-doped $TiSi_2$ is currently available in the literature to the best of the authors knowledge. The hardness value indicated

(a)

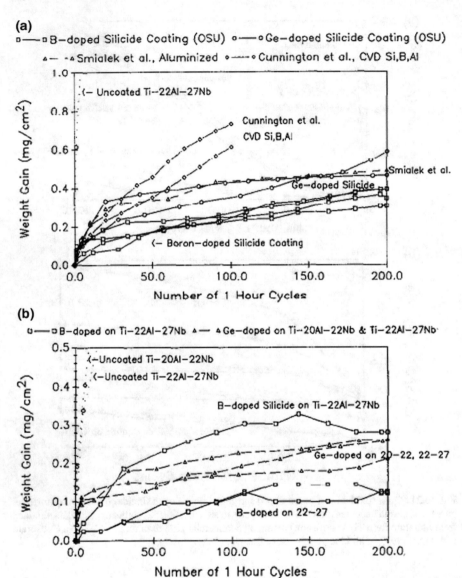

(b)

Fig. 10.22 Cyclic oxidation of boron- and germanium-doped silicide coatings on Ti-22Al-27Nb and uncoated alloys for 200 cycles in air: **a** B-doped (□) and Ge-doped (○) silicide at 1000 °C, compared with published data for (△) an aluminized Ti-14Al-21Nb and (□) a Si-B-Al CVD coating on Ti-14Al-21Nb; **b** B-doped (□) silicide on Ti-22Al-27Nb and Ge-doped (△) silicide on Ti-20Al-22Nb and Ti-22Al-27Nb at 800 °C. Cockeram and Rapp (1995). With kind permission of Elsevier

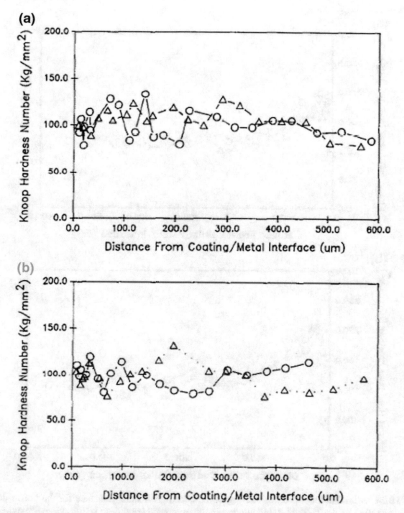

Fig. 10.23 Mocrohardness profiles starting from the coating/substrate interface for the boron-doped silicide coating on commercial pure titanium: **a** in the as-coated condition; **b** following 200 oxidation cycles at 800 °C. Cockeram and Rapp (1995). With kind permission of Elsevier

is an evaluation at the B-doped TiSi$_2$/coating interface. One might be puzzled why such data are lacking despite the large research activity in B-doped TiSi$_2$. The reason is likely due to the fact that all efforts were directed to evaluate its use in microelectronics in particular in VLSI.

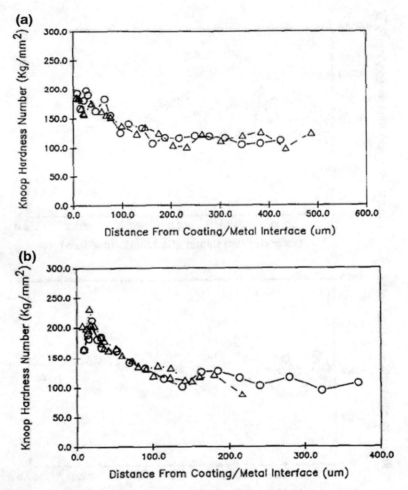

Fig. 10.24 Microhardness profiles starting at the coating/substrate interface for the boron-doped silicide coating on the Ti-22Al-27Nb alloy: **a** in the as-coated condition; **b** following 200 oxidation cycles at 800 °C. Cockeram and Rapp (1995). With kind permission of Elsevier

Summary

- Grain boundaries are strengthened by B
- B addition improves oxidation resistance and eliminates pest formation
- Other mechanical properties such as hardness and fracture toughness are also improved
- The exact mechanism of the B effect still remains controversial.

References

B.V. Cockeram, R.A. Rapp, Mater. Sci. Eng. A **192/193**, 980 (1995)

A. Malchère, P. Gas, C. Haut, A. Larewre, T.T. Nguyen, S. Poize, Appl. Surf. Sci. **53**, 132 (1991)

M. Meyer, M. Kramer, M. Akinc, Adv. Mater. **8**, 85 (1996)

A. Shan, W. Fang, H. Hashimoto, Y.-H. Park, Mater. Trans. **43**, 5 (2002)

P.R. Taleghani, S.R. Bakhshi, M. Erfanmanesh, G.H. Borhani, R. Wafaei, Powder Technol. **254**, 241 (2014)

J.-M. Ting, J. Mater. Sci. Lett. **14**, 539 (1995)

Chapter 11
Silicide Composites

Abstract To improve strength and creep resistance of $MoSi_2$, WSi_2 and $TiSi_2$ reinforcing phases are added to produce silicide composites. Their effect depends on size and shape. Common composites of these silicides are based on addition of SiC, Si_3N_4 and other silicides such as $TaSi_2$. An effective silicide added as a strengthener to $MoSi_2$ is Mo_5Si_3. The low fracture toughness below the ductile to brittle transition temperature in WSi_2 is expected to be improved by SiC additions. The improvement in the fracture toughness in WSi_2 and $TiSi_2$ is a result of crack deflection and crack bridging.

11.1 Introduction

A composite is similar to an alloy, thus a mixture of two or more components. The components are chemically and physically different from one another, and the purpose of their combination is to create a composite that is stronger than the original components would be on their own. Composites are always heterogeneous.

Of the many and various composites based on silicides only a few of known strengtheners such as SiC and Ni_3Si_4 and other silicide additions considered in this book will be discussed. Of the composites the most commonly studied composites will be selected, as a matter of fact one composite for each silicide pending on the availability in the literature. Clearly only silicides composites are chosen which bare connection to the mechanical properties of silicides which is basically the subject of this book.

11.2 MoSi$_2$ Based Composites

Production methods and the components determine the various properties of the composite. In the following this subject is discussed.

© Springer Nature Switzerland AG 2019
J. Pelleg, *Mechanical Properties of Silicon Based Compounds: Silicides*,
Engineering Materials, https://doi.org/10.1007/978-3-030-22598-8_11

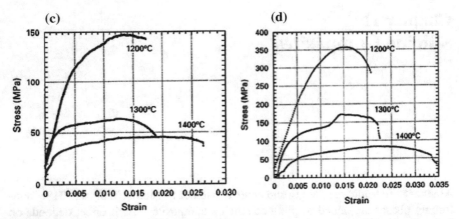

Fig. 11.1 Stress-strain curves for hot pressed samples at all temperatures and strain rates: **c** 10^{-5} s^{-1} and **d** 10^{-5} s^{-1}. Bartlett and Castro (1998). With kind permission of Springer Nature

11.2.1 MoSi₂/SiC

MoSi$_2$ is a high temperature structural material. The ductile to brittle transition temperature is around 1100 °C (Bartlett and Castro) which is the reason for each high temperature toughness. The major limitations for its high temperature use are the reduction of strength and creep resistance. Hard reinforcing phases may provide strength and creep resistance and the effect depends on their amount and size among other factors. A self propagating high temperature synthesis technique has been used for the production Hard reinforcing phases may provide strength and creep resistance and the effect depends on their amount and size among other factors. The stress-strain curves at three temperatures are illustrated in Fig. 11.1 at two strain rates of 10^{-5} s^{-1} for (c) and 10^{-4} s^{-1} for (d). Almost complete densification was the result of the hot pressing of the silicide phase of the MoSi$_2$/SiC powders while considerable agglomeration of the SiC occurred in 20 μm particle sizes. The large agglomerates were evenly distributed throughout the matrix and located on MoSi$_2$ grain boundaries as seen in Fig. 11.2. No crack initiation was observed at the tensile surface which propagated through the sample. The cracking occurred in the SiC but not in the matrix. No interface decohesion was observed. Figure 11.3 shows incipient cracking at 1400 °C in the SiC occurring ahead of many cracks that were present in the sample. The extent of damage zone around the cracks was greater in samples tested at high temperatures compared with those tested at low temperatures. Also the number of cracks was temperature dependent. It is suggested that damage can be minimized by reducing the reinforcement particle, namely, in this case SiC.

11.2.2 MoSi₂/Si₃N₄

Another hard reinforcing component added to MoSi$_2$ in an attempt for increasing its strength is Si$_3$N$_4$. Figure 11.4 illustrates the temperature effect on the stress-

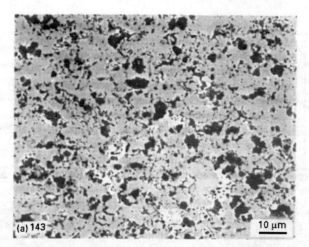

Fig. 11.2 Microstructure of hot-pressed MoSi$_2$/SiC. Dark phase is SiC. Bartlett and Castro (1998). With kind permission of Springer Nature

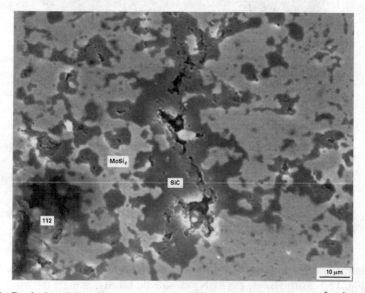

Fig. 11.3 Cracked particles of hot pressed MoSi$_2$/SiC tested at 1400 °C and 10^{-5} s^{-1}. Bartlett and Castro (1998). With kind permission of Springer Nature

strain relation at two strain rates. Figure 11.4 indicates a higher yield stress than the one observed in MoSi$_2$/SiC illustrated in Fig. 11.1, despite the lower volume fraction of the reinforcing Si$_3$N$_4$. The microstructure of the MoSi$_2$/Si$_3$N$_4$ is seen in Fig. 11.5. The agglomerates were located on the MoSi$_2$. Damage by multiple cracks is seen in Fig. 11.6. The damaged zone is ahead of the crack tip. Interface separation occurred in the MoSi$_2$/Si$_3$N$_4$ composites. The damage was more extent at the higher

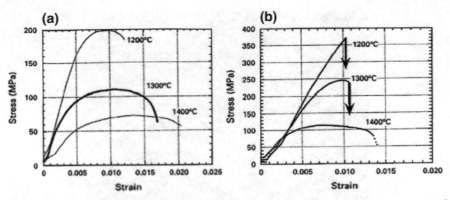

Fig. 11.4 Stress-strain curves for hot pressed samples at all temperatures and strain rates: **a** 10^{-5} s^{-1} and **b** 10^{-4} s^{-1}. Bartlett and Castro (1998). With kind permission of Springer Nature

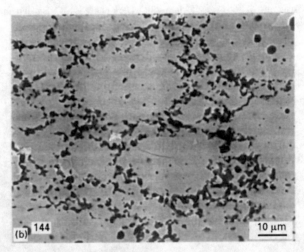

Fig. 11.5 Microstructure of hot-pressed MoSi$_2$/Si$_3$N$_4$. Dark phase is Si$_3$N$_4$. Bartlett and Castro (1998). With kind permission of Springer Nature

temperatures and the major crack path was primarily intergranular/interparticle as seen in Fig. 11.7. Surprisingly the Si$_3$N$_4$ reinforced MoSi$_2$, namely the MoSi$_2$/Si$_3$N$_4$ composite shows a lower stress-strain relation compared to MoSi$_2$ as illustrated for the hot pressed self propagating high temperature synthesis in Fig. 11.8. The ultimate strength and strain to failure of the composite are much reduced compared to the monolithic MoSi$_2$. The type of consolidation influences the stress-strain relation. In Fig. 11.9 plasma sprayed high temperature synthesis technique (SHS) MoSi$_2$/Si$_3$N$_4$ is compared with the monolithic MoSi$_2$. Note that the plasma sprayed composite showed lower yield and ultimate tensile strength compared to hot pressed material and exhibits power law hardening behavior and much greater strain tolerance. The stress-strain curves of the monolithic MoSi$_2$ and the plasma sprayed high temperature

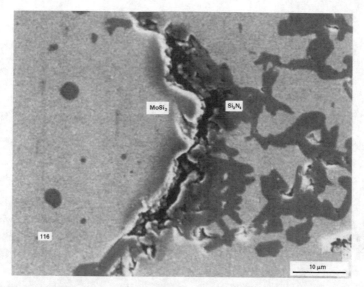

Fig. 11.6 Interfacial separation in the MoSi$_2$/Si$_3$N$_4$ hot pressed powders tested at 1400 °C and 10^{-5} s^{-1}. Bartlett and Castro (1998). With kind permission of Springer Nature

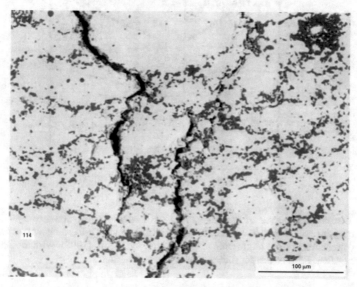

Fig. 11.7 Gross cracking behaviour in the MoSi$_2$/Si$_3$N$_4$ hot-pressed powders, tested at 1400 °C and 10^{-5} s^{-1}. Bartlett and Castro (1998). With kind permission of Springer Nature

synthesis technique tested at 1200 °C and 10^{-5} s^{-1} are presented in Fig. 11.9. The plasma prayed material showed higher yield strength than the sprayed monolithic MoSi$_2$ as seen in Fig. 11.9. Compare Figs. 11.8 and 11.9. It is seen that in the plasma

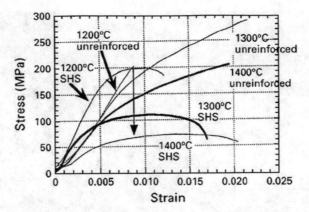

Fig. 11.8 Comparison of the stress-strain response of hot-pressed SHS MoSi$_2$/Si$_3$N$_4$ with mono-lithic hot-pressed MoSi$_2$ at 10^{-4} s^{-1}. Bartlett and Castro (1998). With kind permission of Springer Nature. SHS stand for high temperature synthesis technique

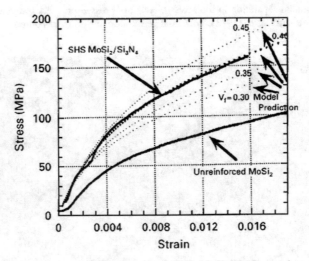

Fig. 11.9 Stress-strain curves of plasma-sprayed SHS MoSi$_2$/Si$_3$N$_4$ powders and monolithic sprayed MoSi$_2$ at 1200 °C and 10^{-5} s^{-1}. FEM analysis of Bao et al. (1991) is shown for comparison at various volume fractions of reinforcing phase. Bartlett and Castro (1998). With kind permission of Springer Nature. FEM stands for finite element analysis method

sprayed material the unreinforced material is at lower stress than the reinforced MoSi$_2$/Si$_3$N$_4$ contrary to the hot pressed case. The increased strength of the plasma sprayed material is attributed to the presence of Mo$_5$Si$_3$ which acts as a reinforcing phase. Also the finer grain size of the plasma sprayed material in addition to the Mo$_5$Si$_3$ precipitate contributes to the increased strengh.

11.3 MoSi₂/Silicide Composite

Of the many MoSi$_2$/Silicide composites three types are considered (i) with TaSi$_2$. (ii) with WSi$_2$ and (iii) with Mo$_5$Si$_3$. MoSi$_2$/Mo$_5$Si$_3$ is an example of a two phase eutectic composite.

(i) MoSi₂/TaSi2

The composite was fabricated from Mo, Ta and Si nanopowders by high energy ball milling and sintered by the pulsed current activated sintering method (PCASM). The sintering temperature of the powder is lower due to the increased reactivity, The grain growth is controlled by the PCASM because dense material can be obtained within a short time already of ~2 min. The grain size was calculated by the method of Suryanarayana and Norton (1998) according to

$$B_r\left(B_{crystalline} + B_{strain}\right)\cos\beta = \frac{k\lambda}{L} + \eta \sin\theta \tag{11.1}$$

where B$_r$ is the full width at half-maximum (FWHM) of the diffraction peak after instrumental correction, B$_{crystalline}$ and B$_{strain}$ are the FWHM caused by grain size and internal stress, respectively, k is a constant with a value of ~0.9, λ is the wave length of the radiation, L and η are the grain size and the internal strain, respectively and θ is the Bragg angle. Further from Cauchy's relation

$$B = B_r + B_s \tag{11.2}$$

where B and B$_s$ are the FWHM of the broadened and the standard Bragg peaks of the samples, respectively.

After milling the powder was placed in a graphite die and then introduced into the pulse current activated sintering system. After evacuation a uniaxial pressure of 80 MPa was applied. Temperatures were measured by a pyrometer. The sequence of experiments can be found in detail in the publication of Park et al. Sintering at 1120 °C resulted in a dense 0.5MoSi$_2$-0.5TaSi$_2$ composite with an average grain size of 60 and 46 nm. A microstructure obtained under these conditions is indicated in Fig. 11.10. Vickers hardness measurements were made on the polished composite specimen using a 1 kg load with a dwell time of 15 s. Figure 11.11a illustrates the hardness indentation and a crack propagating in the composite is seen in Fig. 11.11b. The calculated hardness of the composite is 1320 kg/mm^2 which is an average of five measurements. With a large enough indentation load a median crack develops around the indent as is seen in Fig. 11.11b. From the length of such cracks the fracture toughness can be determined by the expression of Niihara et al. (1982) given as

$$K_{Ic} = 0.023(c/a)^{-3/2}H_V a^{1/2} \tag{11.3}$$

In Eq. (11.3) c is the trace length of the crack measured from the center of the indentation, a is the half of the average length of the two indent diagonals and H$_V$ is

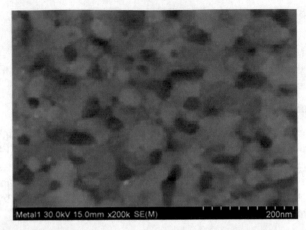

Fig. 11.10 FE-SEM (field emission scanning electron microscope) image of $0.5MoSi_2$-$0.5TaSi_2$ composite sintered at 1120 °C. Park et al. (2013). With kind permission of Springer Nature

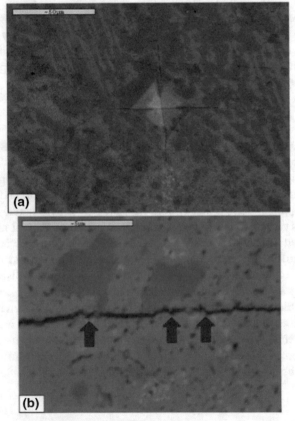

Fig. 11.11 **a** Vickers hardness indentation and **b** median crack propagating in $0.5MoSi_2$-$0.5TaSi_2$ composite. Park et al. (2013). With kind permission of Springer Nature

the Vickers hardness. The toughness determined by this calculation method evaluated from five measurements is 4.1 MPa m$^{1/2}$. These values of hardness and fracture toughness of the nano structured 0.5MoSi$_2$-TaSi$_2$ are higher than the micron structured monolithic MoSi$_2$ (hardness 8.7 MPa and fracture toughness of 2.58 MPa m$^{1/2}$).

(ii) MoSi$_2$/WSi$_2$

Potential use at elevated temperatures of structural materials is improved by WSi$_2$ addition. Monolithic MoSi$_2$ is extremely brittle and shows poor impact strength at lower temperatures and exhibits low strength and creep resistance at temperatures >1200 °C. By adding a second phase MoSi$_2$ can be strengthened and toughened. It is expected that MoSi$_2$ is significantly strengthened by WSi$_2$ addition in particular the nanocrystalline MoSi$_2$/WSi$_2$ properties are enhanced. The specimens were prepared by pressureless sintering from mechanically-assisted combustion synthesis of the powders (Combustion synthesis is also known as self propagating high temperature synthesis). To eliminate fabrication difficulties the composite was prepared by the mechanically assisted self-propagating high temperature synthesis (MASHS) method. The composite showed excellent mechanical properties and good oxidation resistance at the lower temperatures. Table 11.1 shows the mechanical properties of the MoSi$_2$/WSi$_2$ sintered composite. The data of Ref. 22 are from Ai et al. (2005) and prepared by pressureless sintering.

The microstructure of the WMS1 and WMS2 sintered samples illustrating the fractured surfaces is shown in Fig. 11.12.

WSM2 has the better properties as seen in Table 11.1 and they are related to its higher density of 96.49% compared with the other samples indicated in Table 11.1 including the data of Ai et al. (obtained by pressureless sintering) having a lower density. As seen in Fig. 11.12 WSM2 has finer grains than WSM1. Further the higher density of WSM2 is related to the fact that the grains are connected with only independent holes, while in WSM1 the opposite is true (individual grains are still seen) and the holes in large are connected. Thus WSM1 did not sinter completely. Thus the properties of WSM2 as seen in Table 11.1 are better than those of the other samples. The oxidation behavior of WSM2 at 500 °C is seen in Fig. 11.13 expressed as the mass change versus the time of the cyclic oxidation. At the beginning (about

Table 11.1 Mechanical properties of sintered products. Xu et al. (2015). With kind permission of the Japan Institute of Metals and Materials

Samples	Milling conditions for raw materials	Size of as-synthesized powder $D/\mu m$	Density $\rho/g\ cm^{-3}$	Relative density/%	Flexural strength σ_f/MPa	Vieker's hardness HV/GPa	Fracture toughness $K_{1c}/MPa\ m^{1/2}$
WMS1	200 rpm 100 h	1.93	6.10	87.27	210.80	9.72	6.69
WMS2	300 rpm 20 h	0.89	6.74	96.49	327.21	10.78	7.32
WMS3	400 rpm 3 h	1.08	6.37	91.13	242.27	10.56	6.69
MoSi$_2$/WSi$_2^{22}$	–	–	–	90.2	250	8.17	–

Fig. 11.12 SEM
micrographs of sintered
samples: **a** WMS1;
b WMS2. Xu et al. (2015).
With kind permission of the
Japan Institute of Metals and
Materials

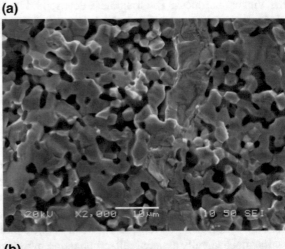

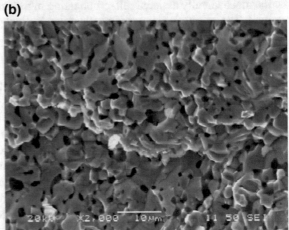

Fig. 11.13 Cyclic oxidation
curve of WMS2 at 500 °C.
Xu et al. (2015). With kind
permission of the Japan
Institute of Metals and
Materials

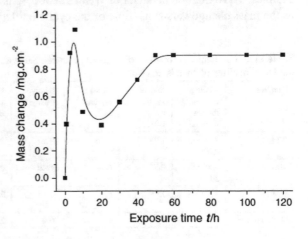

the first 5 h) a rapid increase in the weight occurs. At the this stage surface oxidation occurs and a discontinuous not dense oxide is formed (probably MoO_3, WO_3 and SiO_2). The weight loss is a result of MoO_3 and WO_3 evaporation from the surface. The oxidation continues with the elapse of time and the SiO_2 film acting as a protection layer covers gradually the matrix inhibiting further oxidation when entirely covering the surface. The oxide is stable after ~50 h. A microstructure of the oxidized sample in air for 120 h at 500 °C is illustrated in Fig. 11.14. There are some cracks and some holes in the oxide layer.

In summary the relative density, flexural strength, Vickers hardness and fracture toughness of the sintered $MoSi_2/WSi_2$ are 96.49%, 327.21 MPa, 10.78 GPa and 7.32 MPa m$^{1/2}$, respectively as listed in Table 11.1. Also the composite exhibits a good low temperature oxidation resistance.

(iii) $MoSi_2/Mo_5Si3$

Below the mechanical properties of directionally solidified (DS) $MoSi_2/Mo_5Si_3$ composite is considered. The two-phase eutectic DS ingots were prepared by arc melting and from these ingots the composites were grown using an optical floating zone furnace at various rates in the range of 5–100 mm/h under an argon atmosphere. SEM electron back scattered images are shown in Fig. 11.15. Table 11.2 lists pertinent data covering the $MoSi_2/WSi_2$ composite. There are some uncovered areas for example area A, and other areas such as B where the amount of oxygen is less as indicated by EDS results of the $MoSi_2$ and Mo_5Si_3 components of the composite. Dark and bright regions are observed in the images of Fig. 11.15 which correspond respectively to $MoSi_2$ and Mo_5Si_3. The SEM backscattered electron images seen in Fig. 11.15 are a cross-section of the DS eutectic ingot (($1\bar{1}0)_{MoSi_2}//(001)_{Mo_5Si_3}$). The grown in dislocation structure of the DS eutectic grown at 100 mm/h is seen in Fig. 11.16. The dislocation structure is mainly concentrated in the $MoSi_2$ matrix as illustrated in the TEM thin film cut parallel to the orientation indicated above. The Burgers vector of the grown in dislocations is ⟨100⟩ evaluated by TEM contrast analysis. It is believed

Fig. 11.14 SEM micrograph of oxidized WMS2 after exposure to air for 120 h. Xu et al. (2015). With kind permission of the Japan Institute of Metals and Materials

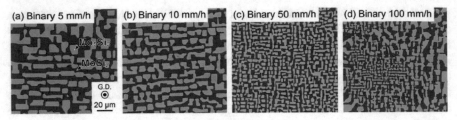

Fig. 11.15 SEM backscattered electron images of DS ingots of **a** 5 mm h^{-1}, **b** 10 mm h^{-1}, **c** 50 mm h^{-1} and **d** 100 mm h^{-1}. Matsunoshita et al. (2016). Taylor and Francis publication, open access of the creative common attribution licence (http://creativecommons.org/licenses/by/4.0/)

that the dislocations were introduced during cooling as a result of the mismatch of the thermal expansion coefficients between the two phases.

Stress strain curves of the $[001]_{MoSi_2}$ oriented binary DS composites are illustrated in Fig. 11.17. Plastic flow is observed only above 1000 °C and below it it is brittle. The stress strain curves at various growth rates can be seen for the $[1\bar{1}0]_{MoSi_2}$ oriented specimens in Fig. 11.18. As seen some ternary composites are also included. The yield stress variation with temperature is shown in Fig. 11.19. Clearly as expected the yield stress decreases with temperature increase, but the important observation is that the stress of the DS eutectic composite is much higher than the single crystals of the monolithic MoSi$_2$ (Fig. 11.19a) and about the same as the Mo$_5$Si$_3$ single crystal (Fig. 11.19b) but higher in the temperature range 1300–1400 °C. A TEM bright field image of $[110]_{MoSi_2}$—oriented composite specimen deformed to 1% plastic strain at 1000 °C is seen in Fig. 11.20. Dislocation are observed in both phases of the DS eutectic composite. The Burgers vectors of the dislocations in the MoSi$_2$ matrix are [100] and [010] designated as **b**$_1$ and **b**$_2$ respectively and are identical to those of as-grown dislocations (Fig. 11.16). Activation of ⟨100⟩ dislocations are similarly observed in single crystal MoSi$_2$ oriented at $[1\bar{1}0]_{MoSi_2}$. Plastic flow in Mo$_5$Si$_3$ is at or above 1300 °C and the introduction of dislocations with the [001] orientation is assisted by the stress concentration generated by dislocation pile ups against the interface boundary in the MoSi$_2$ matrix. At some critical value the plastic deformation initiated at the MoSi$_2$ matrix propagates into the Mo$_5$Si$_3$. The TEM images of the DS eutectic composites deformed at 1000 and 1400 °C are illustrated in Fig. 11.20.

Contrast analysis shown in Fig. 11.21 illustrates the dislocation introduced to the Mo$_5$Si$_3$ at 1000 °C. Dislocations with three Burgers vectors exist in Mo$_5$Si$_3$ marked A, B and C. Dislocations marked A are invisible when the diffraction vectors, **g** are $0\bar{2}2$ (Fig. 11.21d) and 202 (Fig. 11.21f). According to the **g·b** rule, the Burgers vector **b**$_A$ of dislocation A is determined to be parallel to $[1\bar{1}\bar{1}]$. The Burgers vector **b**$_B$ of dislocation B is determined to be parallel to $[1\bar{1}1]$ since it is invisible to **g** $\bar{1}12$ (Fig. 11.21c) and **g** 022 (Fig. 11.21e). For Mo$_5$Si$_3$ (body-centered tetragonal crystal structure of the D8 type: space group I4mcm) the shortest translation vector along ⟨111⟩is 1/2⟨111⟩. Thus the Burgers vectors of the observed dislocations A and B are inferred to be 1/2$[1\bar{1}\bar{1}]$ and 1/2$[1\bar{1}1]$, respectively. Dislocations C are invisible for $\bar{3}30$ (Fig. 11.21b) and are visible for the other imaging conditions (Fig. 11.21). Of

Table 11.2 Average thickness, Lamellar spacing and chemical composition of the MoSi$_2$ and Mo$_5$Si$_3$ phases in the binary DS eutectic composites grown at various growth rates. Matsunoshita et al. (2016). Taylor and Francis publication, open access of the creative common attribution licence (http://creativecommons.org/licenses/by/4.0/)

Growth rate (mm h^{-1})	Average thickness (µm)		Average lamellar spacing (µm)	Chemical composition				
				MoSi$_2$			Mo$_5$Si$_3$	
	MoSi$_2$	Mo$_5$Si$_3$		Mo	Si	x	Mo	Si
Binary 5	9±8	9±6	18	35.2	64.8	–	63.2	36.8
10	7±6	7±4	14					
55	3±2	3±1	6					
100	4±3	4±2	8					
(Cell center)	3±2	3±2	6					
(Cell boundary)	13±10	13±8	26					

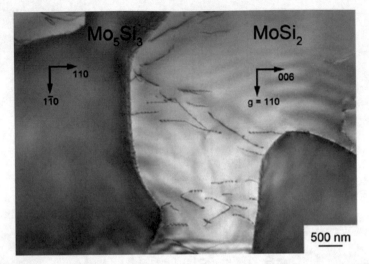

Fig. 11.16 Growth-in dislocations in a binary DS eutectic composite grown at a rate of 100 mm h^{-1}. Matsunoshita et al. (2016). Taylor and Francis publication, open access of the creative common attribution licence (http://creativecommons.org/licenses/by/4.0/)

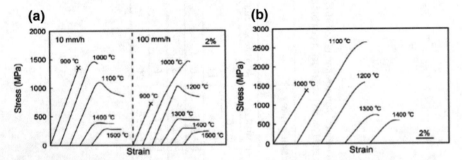

Fig. 11.17 Stress-strain curves of binary DS eutectic composites tested in compression [1$\bar{1}$0]$_{MoSi_2}$ and **b** [001]$_{MoSi_2}$. Matsunoshita et al. (2016). Taylor and Francis publication, open access of the creative common attribution licence (http://creativecommons.org/licenses/by/4.0)

the possible vectors (i.e. [001], 1/2$\langle 111 \rangle$, $\langle 100 \rangle$ and $\langle 011 \rangle$) only **b** = [001] can satisfy the visibility conditions. The Burgers vector **b**$_C$ of dislocation C is thus determined as [001].

   In Mo$_5$Si$_3$ the dislocation A and B (**b** = 1/2$\langle 111 \rangle$, i.e. for **b**$_A$ and **b**$_B$) can glide on their appropriate slip planes under the influence of external stress, but dislocation C (**b**$_C$ = [001]) cannot glide easily because no shear stress is acting on it. The slip planes for dislocations A and B (**b**$_A$ and **b**$_B$) are ($\bar{1}$12) and (1$\bar{1}$2). There are three different Burgers vectors in Mo$_5$Si$_3$ marked as A, B and C and given as 1/2[111], ½[11$\bar{1}$] and [001] and are obtained by contrast analysis. There is no shear stress acting along [001] with [110]$_{Mo_5Si_3}$ loading axis orientation, therefore, the dislocation C (**b** = [001]) is likely to be the result of dislocation reactions between 1/2$\langle 111 \rangle$.

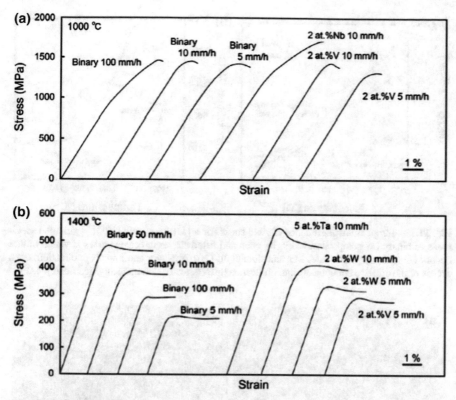

Fig. 11.18 Stress-strain curves obtained for $[1\bar{1}0]_{\text{MoSi}_2}$—oriented specimens and some ternary eutectic composites at **a** 1000 and **b** 1400 °C. Matsunoshita et al. (2016). Taylor and Francis publication, open access of the creative common attribution licence (http://creativecommons.org/licenses/by/4.0/)

Thus, the plastic flow of the eutectic composite sets in above 1000 °C when the loading axis is parallel to $[1\bar{1}0]_{\text{MoSi}_2}$, while when it is parallel to $[001]_{\text{MoSi}_2}$ it occurs above 1100 °C. In binary DS eutectic oriented in $[1\bar{1}0]_{\text{MoSi}_2}$ the yield stress is about 3–10 times higher than that of MoSi$_2$ single crystal oriented at [110] (see Fig. 11.19a) in the temperature range 1000–1500 °C. At $[001]_{\text{MoSi}_2}$ orientation however (Fig. 11.19b) it exhibits yield stress values higher than the respective MoSi$_2$ single crystal at 1100 °C, but the opposite is true at temperatures above 1300 °C. The yield stress of $[110]_{\text{MoSi}_2}$ oriented DS MoSi$_2$/Mo$_5$Si$_3$ eutectic composites depends on the growth rate during the DS solidification with the plastic flow initiated in the MoSi$_2$ matrix. Therefore there is a size effect of the MoSi$_2$ matrix, namely its thickness. There is an increase in yield stress with the average decrease of the thickness of the MoSi$_2$ phase in the composite. For more details-in particular for the strain tensor—the reader is referred to the publication of Matsunoshita et al. (2016).

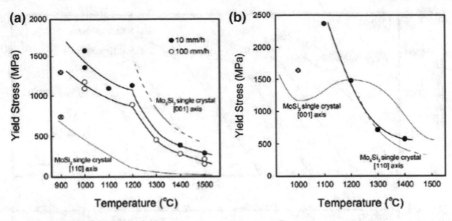

Fig. 11.19 Temperature dependence of yield stress for **a** $[1\bar{1}0]_{MoSi_2}$ and **b** $[001]$—oriented specimens of binary DS composites. Marks by open and filled circles indicate stresses at which failure occurs without any plastic flow. Matsunoshita et al. (2016). Taylor and Francis publication, open access of the creative common attribution licence (http://creativecommons.org/licenses/by/4.0/)

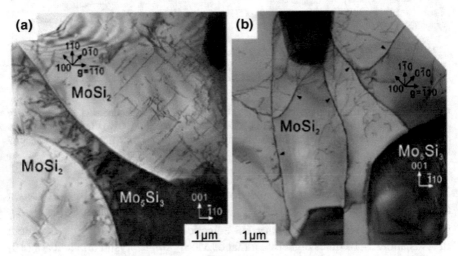

Fig. 11.20 TEM bright-field images of $[1\bar{1}0]_{MoSi_2}$—oriented specimens of binary DS eutectic composites deformed at **a** 1000 °C and **b** 1400 °C. Matsunoshita et al. (2016). Taylor and Francis publication, open access of the creative common attribution licence (http://creativecommons.org/licenses/by/4.0/)

11.4   WSi$_2$ Based Composites

(a)   WSi$_2$/SiC

WSi$_2$/SiC composite was prepared from WC and Si powders by high frequency induction heated combustion synthesis (HFIHCS). This fabrication combines the

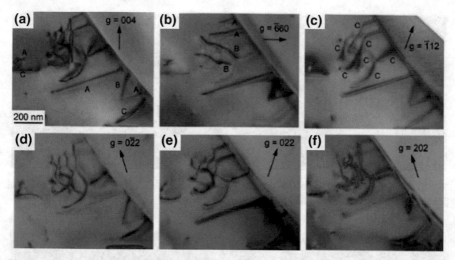

Fig. 11.21 Contrast analysis of dislocations in a $[1\bar{1}0]_{MoSi_2}$—oriented specimen of a binary DS eutectic composite deformed at 1000 °C. The diffraction vector (**g**) used is indicated in each of the images. Matsunoshita et al. (2016). Taylor and Francis publication, open access of the creative common attribution licence (http://creativecommons.org/licenses/by/4.0/)

effect of an induced current and mechanical pressure (60 MPa) resulting in high density (up to 97%) WSi$_2$/SiC in a one step process. The average grain sizes of WSi$_2$ and SiC were 0.78 and 0.56 μm respectively. The resulting hardness and fracture toughness values are 1790 kg/mm^2 and 4.4 MPa m$^{1/2}$, respectively. High temperature and electronic applications of such composites are of interest for potential use. Both components have excellent creep and oxidation resistance and WSi$_2$ also has high melting point (2160 °C) and good strength at elevated temperatures. As many intermetallic compounds, WSi$_2$ has low fracture toughness below the ductile to brittle transition temperature and therefore it is expected that as a composite with SiC the mechanical properties might improve.

The starting raw materials are seen in Fig. 11.22. The microstructures at various stages of WSi$_2$/SiC composite fabrication is illustrated in Fig. 11.23. Vickers hardness indentation is illustrated in Fig. 11.24a and the resulting crack propagation in the composite is seen in Fig. 11.24b. Densification during fabrication by the high-frequency induction heated combustion synthesis is shown in Fig. 11.25 and the mapping of combustion synthesized WSi$_2$-SiC composite is presented in Fig. 11.26. The average grain size was evaluated by the linear intercept method. The mean intercept length l for the WSi$_2$, l$_{WSi_2}$ and for the SiC, l$_{SiC}$ was evaluated by counting the number of each grain, i.e. N$_{SiC}$ and N$_{WSi_2}$ intercept randomly and using the relations:

$$\bar{l}_{WSi_2} = \frac{V_{WSi_2}(L/M)}{N_{WSi_2}}, \quad \bar{l}_{SiC} = \frac{V_{SiC}[L/M]}{N_{SiC}} \tag{11.4}$$

Fig. 11.22 Scanning electron microscope images of raw materials: **a** tungsten carbide, **b** silicon powder. Oh et al. (2005). With kind permission of Elsevier

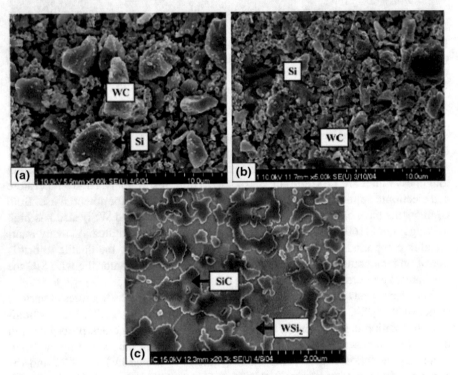

Fig. 11.23 Scanning electron microscope images of WC + 3Si system: **a** after milling, **b** before combustion synthesis, **c** after combustion synthesis. Oh et al. (2005). With kind permission of Elsevier

where the V's represent the volume fractions of the respective components in the composite expressed as fraction, the L's the test lines lengths and the M's are the magnification. The average grain size was obtained from

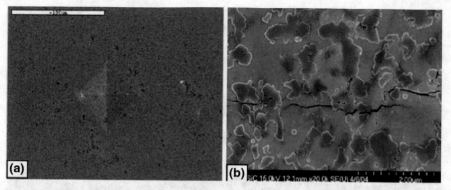

Fig. 11.24 **a** Vickers hardness indentation and **b** median crack propagating of WSi₂-SiC composite. Oh et al. (2005). With kind permission of Elsevier

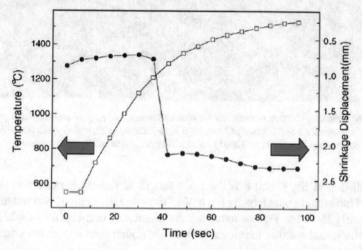

Fig. 11.25 Variations of temperature and shrinkage displacement with heating time during high-frequency induction heated combustion synthesis and densification of WSi₂-SiC composite. Oh et al. (2005). With kind permission of Elsevier

$$\bar{D} = 1.571\bar{l} \qquad (11.5)$$

The average grain sizes obtained by the HFIHCS method are 0.78 μm and 0.56 μm. As indicated above the calculated Vickers hardness value obtained is 1790 kg/mm². The hardness measurement experiments were done under a load of 10 kg and a dwell time of 15 s. The hardness value represents an average of five measurements. With sufficient high loads median cracks were obtained which enabled the evaluation of the fracture toughness by (Eq. 11.3) which is rewritten here as

$$K_{Ic} = 0.023(c/a)^{-3/2}H_v a^{1/2} \qquad (11.3)$$

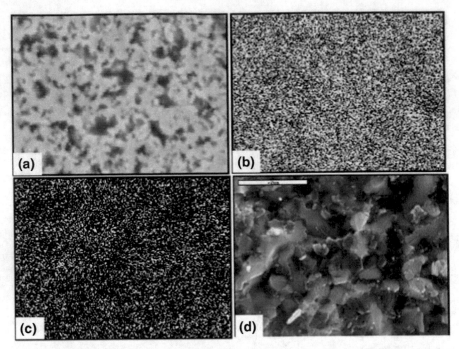

Fig. 11.26 Scanning electron microscope images and X-ray mapping of combustion synthesized WSi$_2$-SiC composite: **a** SEM image of product, **b** X-ray mapping: Si, **c** X-ray mapping: W, **d** SEM image of fracture surface. Oh et al. (2005). With kind permission of Elsevier

Recall that in Eq. (11.3) c is the trace length of the crack measured from the center of indentation with a being the half of the average length of two indent diagonals. Clearly H$_V$ is the Vickers hardness. As mentioned earlier K$_{Ic}$ is 4.4 MPa m$^{1/2}$. The hardness and fracture toughness values are higher than any of the composite's components.

(b) **WSi$_2$/Silicides**

(1) **WSi$_2$/TaSi$_2$**

WSi$_2$/TaSi$_2$ composite was fabricated by using simultaneously 80 MPa pressure and a pulsed current activated sintering method (PCASM). 2 min is sufficient to produce a dense composite. Due to the rapid process by this method, almost a theoretically dense material is obtained with grain growth being retarded. Thus allowing to produce from the nano powders, a dense nanostructure composite. The grain size was calculated using Suryanarayana and Norton's formula which we have seen earlier and is rewritten as:

$$B_r\left(B_{crystalline} + B_{strain}\right)\cos\beta = \frac{k\lambda}{L} + \eta\sin\theta \tag{11.1}$$

Recall that B$_r$ is the full width at half maximum (FWHM) of the diffraction peak after instrumental correction, B$_{crystalline}$ and B$_{strain}$ are the FWHM caused by the grain size and internal stress, k is a constant of 0.9, λ is the wave length of X-ray radiation, L η are the grain size and the internal strain and θ is the Bragg angle. B and B$_r$ follow the Cauchy relation with B = B$_r$ + B$_S$ where B and B$_S$ are the FWHM of the broadened Bragg peaks and the standard Bragg peaks of the samples, respectively.

SEM images of the raw material is shown in Fig. 11.27. Figure 11.28 shows a SEM image of milled powders, and EDS analysis. The grains are very fine, but some agglomerations were observed. Upon heating during densification and the synthesis the change in shrinkage displacement with temperature and time can be seen in Fig. 11.29. A 1200 °C sintered composite under 80 MPa is shown in Fig. 11.30, which is a FE-SEM image of etched sample surface. The microstructure consists of nanophases. The Vickers hardness indent is illustrated in Fig. 11.31a, while in Fig. 11.31b median crack propagation is shown. The hardness measurements were done under a load of 1 kg and a 15 s dwell time. The calculated hardness value of 1190 kg/mm^2 represent the average of five measurements. With sufficient load median cracks propagated around the indentation as seen in Fig. 11.31b. As mentioned in an earlier section the length of these cracks permits the calculation of an estimated fracture toughness by the relation shown earlier, i.e., Equation (11.3). The young modulus appearing in Eq. (11.3) was calculated from the rule of mixture

Fig. 11.27 SEM images of raw materials: **a** W, **b** Ta, and **c** Si. Kim et al. (2014). With kind permission of Elsevier

Fig. 11.28 SEM image (**a**), and EDS (**b**) of the milled 0.5W-0.5Ta-2Si powder. Kim et al. (2014). With kind permission of Elsevier

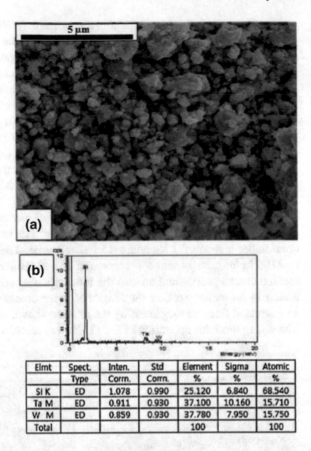

Elmt	Spect. Type	Inten. Corrn.	Std Corrn.	Element %	Sigma %	Atomic %
Si K	ED	1.078	0.990	25.120	6.840	68.540
Ta M	ED	0.911	0.930	37.100	10.160	15.710
W M	ED	0.859	0.930	37.780	7.950	15.750
Total				100		100

Fig. 11.29 Variations of temperature and shrinkage displacement with heating time during synthesis and densification of the 0.5WSi$_2$-0.5TaSi$_2$ composite. Kim et al. (2014). With kind permission of Elsevier

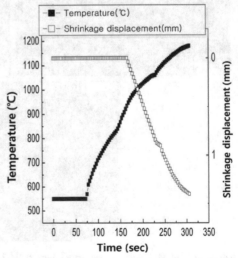

Fig. 11.30   FE-SEM image of 0.5WSi$_2$-0.5TaSi$_2$ composite sintered at 1200 °C. Kim et al. (2014). With kind permission of Elsevier

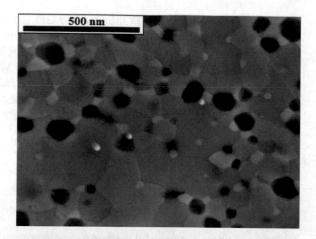

Fig. 11.31   **a** Vickers hardness indentation and **b** median crack propagation in the 0.5WSi$_2$-0.5TaSi$_2$ composite. Kim et al. (2014). With kind permission of Elsevier

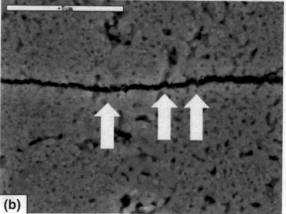

using $E(WSi_2) = 468$ GPa and $E(TaSi_2) = 357$ GPa (for the 0.503 volume fraction of WSi_2 and the 0.497 volume fraction of $TaSi_2$). The resulting fracture toughness value obtained from five measurements is 5 MPa m$^{1/2}$, which is higher than each component of the composite. The fracture toughness values of WSi_2 and $TaSi_2$, respectively are 3.2 and 4.7 MPa m$^{1/2}$. The hardness values of the individual components of the nanostructured composite are, respectively 8.2 MPa and 908 kg/mm^2. One can see from Fig. 11.31a that the cracks propagate from the indent corners.

11.5   TiSi$_2$ Based Composites

(a)   TiSi$_2$/SiC

As indicated above a successful method to produce composites in a one step process in a relatively short period of time of 2 min is the high frequency induced heated combustion synthesis (HFIHCS) which also densifies the materials to obtain a dense product. The basis of the solid state reaction to obtain the TiSi$_2$/SiC composite is

$$TiC + 3Si \rightarrow TiSi_2 + SiC \tag{11.4}$$

A mixture of TiC and Si powders were at first milled in a high-energy ball mill (by tungsten carbide balls) for about 10 h, then sealed in a cylindrical stainless steel vial under argon atmosphere. The milling resulted in small grain size after which the mixed powder was placed in a graphite die and introduced into the high frequency-induced heated combustion system for completion of the fabrication process. Figure 11.32 is a SEM image showing the raw materials used. The shrinkage displacement during the densification as a function of time and at the temperatures applied is illustrated in Fig. 11.33. The shrinkage displacement increased gradually with time and with (temperature). No reaction took place at heating under a pressure

Fig. 11.32 SEM images of raw materials: **a** titanium carbide, **b** silicon powder. Shon et al. (2007). With kind permission of Elsevier

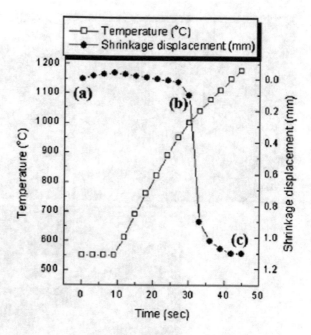

Fig. 11.33 Variations of temperature and shrinkage displacement with heating time during HFIHCS and densification of TiSi$_2$-SiC composite (under 60 MPa, 90% output of total power capacity). Shon et al. (2007). With kind permission of Elsevier

of 60 MPa to 900 °C and there was no significant shrinkage displacement as indicated by XRD and SEM analysis.

The powders reacted when the temperature was raised to 1200 °C producing a highly dense product. SEM image of samples heated to 1200 °C under a pressure of 60 MPa is shown in Fig. 11.34c. This figure shows all stages of fabrication from the powders after milling up to the final composite product of TiSi$_2$/SiC. XRD analysis supports the production result, i.e., the presence of the two phases of TiSi$_2$ and SiC. Backscattered SEM image of polished surface of TiSi$_2$/SiC is seen in Fig. 11.35. The dark areas represent SiC particles which are distributed uniformly in the composite. TEM micrograph of the composite showing the morphology is presented in Fig. 11.36. The grain size of the SiC grains are in the range 200–300 nm. The grains of SiC are mainly distributed in the TiSi$_2$ grain boundaries and strain stripes can be seen because of the large thermal mismatch between the components of the composite. The thermal expansion of TiSi$_2$ is $\alpha_{TiSi_2} = 10 \times 10^{-6}$ K^{-1} and that of SiC $\alpha_{SiC} = 4.8 \times 10^{-6}$ K^{-1}. The mechanical properties of this composite are: Vickers microhardness H$_V$ = 11.5 GPa, fracture toughness, K$_{Ic}$ = 3.3 ± 0.2 MPa m$^{1/2}$ and the bending strength, σ_b = 400 ± 50 MPa. The improvement in the fracture toughness is associated with the combination of the crack deflection and crack bridging. A crack on encountering a SiC particle is partially deflected and part of it propagates through the SiC particle. The crack tends to bypass the SiC particle and propagate along the TiSi$_2$/SiC interface.

In another TiSi$_2$/SiC (Shon et al.) composite discussed earlier the average grain sizes of TiSi$_2$ and SiC were calculated by the Stokes and Wilson formula as indicated

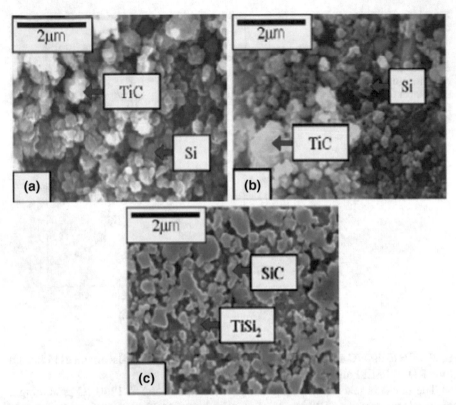

Fig. 11.34 SEM images of the TiC + 3Si system: **a** after milling, **b** before combustion synthesis, **c** after combustion synthesis. Shon et al. (2007). With kind permission of Elsevier

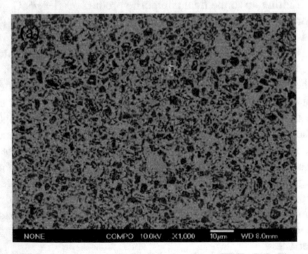

Fig. 11.35 Back scattered electron image of polished surface of $TiSi_2$-SiC. Qin et al. (2006). With kind permission of Japan Institute of Metals

Fig. 11.36 TEM
micrograph of TiSi$_2$-SiC
composites showing the
morphology of TiSi$_2$ and
SiC. Qin et al. (2006). With
kind permission of Japan
Institute of Metals

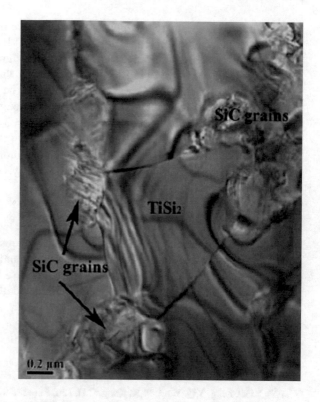

above in Eq. (11.1) by Suryanarayma and Norton's relation. However the completely
equivalent relation of Stokes and Wilson is presented in Eq. (11.5) as

$$b = b_d + b_c = k\lambda/(d\cos\theta) + 4\varepsilon\tan\theta \qquad (11.5)$$

The values are 75 and 76 nm for the TiSi$_2$ and SiC phases, respectively. SiC particles
were well distributed in the matrix. The parameters are equivalent to those in Eq. 11.1,
but nevertheless they are repeated here. The FWHM of the diffraction peak b is a
value after instrument correction. The parameters b_d and b_c are the FWHMs due
to the grain size and the internal stress, respectively The constant k of 0.9 is the
wavelength of the X-ray radiation. Clearly d and ε are the grain size and the strain
by the internal stress, respectively, while θ is the Bragg angle. As mentioned above
from Cauchy's relation

$$B_0 = b + b_s$$

where B_0 and b_s represent the FWHM of broadened Bragg peaks and the standard samples, respectively (see (Eq. 11.2)). The Vickers hardness measurements
were made on polished sections of the TiSi$_2$/SiC composite using a 20 kgf load
and 15 s dwell time. The calculated hardness value of the TiSi$_2$/SiC composite is

1250 kg/mm^2. This value represents an average of five measurements. Sufficiently high loads produced median cracks around the indent. The length of the cracks measured from its center permits the estimation of the fracture toughness by Eq. (11.3). The Young modulus of Eq. (11.3) was estimated by the rule of mixtures for the volume fractions of 0.342 SiC and 0.658 of TiSi$_2$. The values are E(SiC) = 448 GPa and E(TiSi$_2$) = 255.6. The calculated fracture toughness of the composite is ~ 3 MPa m$^{1/2}$. This value is about the same as K$_{Ic}$ = 3.3 ± 0.2 MPa m$^{1/2}$ for the larger grain size TiSi$_2$/SiC composite (200–300 nm) mentioned earlier. The individual fracture toughnesses of the components of the composite are lower, for SiC 1.8 MPa m$^{1/2}$ and for TiSi$_2$ 1.9 MPa m$^{1/2}$, respectively. The results are the average of five measurements.

Summary

- Hard reinforcing phases provide strength and creep resistance
- Improvement by the additives depend on their size and shape
- Additives such as SiC, S$_3$N$_4$ or other silicides improve overall strength properties.
- The effect of the additives is associated with crack deflection and bridging.

References

Y.L. Ai, Y.G. Cheng, Y.Q. Yang, M.K. Kang, C.H. Liu, Rare Metal Mat. Eng. **34**, 962 (2005)
G. Bao, J.W. Hutchinson, R.M. McMeeking, Acta Metall. Mater. **39**, 1871 (1991)
A.H. Bartlett, R.G. Castro, J. Mater. Sci. **33**, 1653 (1998)
D.-K. Kim, I.-J. Shon, Mater. Res. Bull. **57**, 243 (2014)
H. Matsunoshita, Y. Sasai, K. Fujiwara, K. Kishida, H. Inui, Sci. Tech. of Adv. Mater. **17**, 517 (2016)
K. Niihara, R. Morena, D.P.H. Hasselman, J. Mater. Sci. Lett. **1**, 13 (1982)
D.Y. Oh, H.C. Kim, J.K. Yoon, I.J. Shon, J. Alloys Compd. **386**, 270 (2005)
N.-R. Park, J.-M. Doh, J.-K. Yoon, H.-K. Park, I.-J. Shon, Res. Chem. Intermed. **39**, 1269 (2013)
C. Qin, L. Wang, W. Jiang, S. Bai, L. Chen, Mater. Trans. **47**, 845 (2006)
I.-J. Shon, H.-K. Park, H.-C. Kim, J.-K. Yoon, K.-T. Hong, I.-Y. Koa, Scr. Mater. **56**, 665 (2007)
C. Suryanarayana, M. Grant Norton, in *X-Ray Diffraction: A Practical Approach* (Springer, 1998)
J. Xu, Y. Wang, B. Weng, F. Chen, Mater. Trans. **56**, 313 (2015)

Chapter 12
Alloying in Silicides

Abstract Selected alloys of the silicides $NiSi_2$, $MoSi_2$, $FeSi_2$, $TiSi_2$ and Ti_5Si_3 are considered in this chapter. Alloys are either homogeneous or heterogeneous, while composites are always heterogeneous. Metallic elements such as Nb, Al and Cr are additives to improve properties some of them by solid solution strengthening. The room temperature brittle $MoSi_2$ having exceptional oxidation resistance but low high temperature strength need to be improved by reinforcement. Nb or Cr are examples of additives to strengthen $MoSi_2$ and making it a potential material for high temperature applications. Strengthening in the presence of defect complexes occurs either by spherical or nonspherical strain fields. Non-spherical strain fields have a stronger interaction with both edge and screw dislocations than spherical strain fields. In $MoSi_2$ alloyed with Nb non-spherical strain fields may arise. The lack of information of $TiSi_2$ and due to the importance of Ti based materials for high temperature applications unalloyed Ti_5Si_3 and alloyed with Nb and Cr are discussed in the chapter.

12.1 Introduction

At least two components of different properties of a structure make up an alloy. The alloys often have significantly different properties than the components comprising it. In the earlier chapter the mixture of two or more components were also considered but referred to as composites. One can distinguish between the concepts by the character of the structure obtained from the two or more components. An alloy can be either a homogeneous or a heterogeneous mixture of its components, while a composite is always heterogeneous. Of the indefinite possibilities of alloying only selected alloys of the silicides will be considered in this book. Recall that the basic alloys considered in this book are $CoSi_2$, $NiSi_2$, $MoSi_2$, WSi_2, $FeSi_2$ and $TiSi_2$ and therefore alloying of these silicides is of our interest. The alloying components to be chosen for the presentation are those either which are in common use or such that introduce properly changes in the basic alloys. However availability of data in literature clearly limits the type and the content of sections. Thus, no relevant data on alloying of $CoSi_2$ could be found.

© Springer Nature Switzerland AG 2019

J. Pelleg, *Mechanical Properties of Silicon Based Compounds: Silicides*,
Engineering Materials, https://doi.org/10.1007/978-3-030-22598-8_12

12.2 Alloying of NiSi$_2$

Here the effect of another silicide on the unalloyed NiSi$_2$ is discussed. Any additive can be either in the form of solid solution or as a precipitate when the solubility limit is exceeded. Below solid solution of the nickel and iron silicide is presented with the consequent change in the hardness values. Further, the lattice parameter is expected to change on alloying with FeSi$_2$. The overall formula of NiSi$_2$ in the Ni-Si system had a two phase structure as illustrated in Fig. 12.1, a higher nickel-silicide phase known as ξ which contains considerable amounts of silicon crystal.

The alloys with 47.1% Si having the composition Ni$_{1.04}$Si$_{1.93}$ a formula close to NiSi$_2$ had a single-phase structure. The compositions of the alloys is seen in Table 12.1. It is surprising that the lattice parameter decreases with addition of FeSi$_2$ because the radii of the iron atoms (under conditions of identical coordination) is greater than that of the nickel atoms, i.e., $r_{Fe} > r_{Ni}$.

The following compositions of the Ni–Si–Fe were used to evaluate certain properties such as density and the number of atoms of the components NiSi$_2$–FeSi$_2$, Ni$_{1.04}$Si$_{t.93}$–FeSi$_2$, and Ni$_{1.04}$ Si$_{1.93}$–Fe$_{1.04}$Si$_{1.03}$. The specimens with smaller amounts of iron disilicide (5–20% FeSi$_2$) contained in addition to the ξ phase small amounts of crystals of alloyed nickel monosilicide, while those with larger concentrations of the disilicide (40% FeSi$_2$) revealed also the presence of precipitates of alloyed iron disilicide. In Fig. 12.2 the lattice parameter of the ξ phase as a function of composition passes through a minimum. For all the three compositions up to 20% FeSi$_2$ a sharp decrease of the lattice parameters occurs but in specimen (a) beyond 25% a small increase is observed.

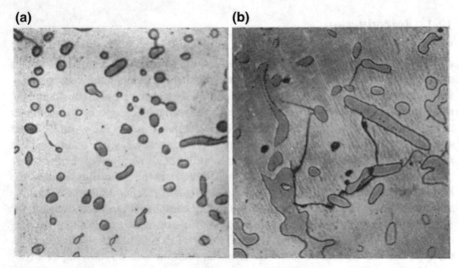

(a)   **(b)**

Fig. 12.1 Photomicrographs of alloys represented by over-all formulas: **a** NiSi$_2$; **b** 0.8 NiSi$_2$ + 0.2 FeSi$_2$. Sidorenko et al. (1968). With kind permission of Springer Nature

Table 12.1 Effect of composition of single-phase Ni$_{1.04}$Si$_{1.93}$ base solid solution on crystalline-lattice parameters (a), density (ρ), number of atoms of components (n$_i$), and number of outer electrons (n$_e$). Sidorenko et al. (1968). With kind permission of Springer Nature

Alloy composition	a, Å	ϱ, g/cm^3	$n_{Me} = n_{Ni} + n_{Fe}$	n_{Si}	$n_{\Sigma} = n_{Me} + n_{Si}$	n_e
100% Ni$_{1.04}$Si$_{1.93}$	5.4066	4.84	4.16	7.73	11.89	72.52
25% FeSi$_2$ + 75% Ni$_{1.04}$ Si$_{1.93}$	5.3916	4.91	4.16	7.89	12.05	71.08
30% FeSi$_2$ + 70% Ni$_{1.04}$ Si$_{1.93}$	5.3920	4.91	4.16	7.92	12.08	70.80
35% FeSi$_2$ + 65% Ni$_{1.04}$Si$_{1.93}$	5.3929	4.91	4.16	7.95	12.11	70.50
30% Fe$_{1.04}$ Si$_{1.93}$ + 70% Ni$_{1.04}$ Si$_{1.93}$	5.3938	4.96	4.27	7.89	12.16	71.66

Fig. 12.2 Effect of concentration of higher iron silicide in three-component system Ni–Si–Fe on crystalline-lattice parameters of ξ phase present in quasi-binary systems:
a Ni$_{1.04}$Si$_{1.98}$–Fe$_{1.04}$Si$_{1.93}$;
b Ni$_{1.04}$Si$_{1.93}$FeSi$_2$;
c NiSi$_2$–FeSi$_2$. Sidorenko et al. (1968). With kind permission of Springer Nature

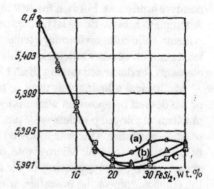

The structural characteristics of solid solutions at the higher nickel and iron silicides are reflected in the properties also as indicated for example in the variation of the microhardness values in Fig. 12.3. Also the resistivity and the thermal expansion are shown of alloys of the system Ni$_{1.04}$Si$_{1.98}$–FeSi$_2$ at 20 °C. The microhardness is a minimum at ~20% FeSi$_2$.

12.3 Alloying of MoSi$_2$

12.3.1 Metallic Element Addition

One of the primary objectives for high temperature application of MoSi$_2$—which has exceptional high temperature oxidation resistance—is to overcome his room tem-

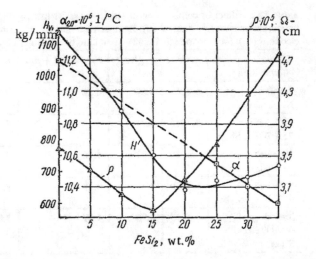

Fig. 12.3 Isotherms of microhardness (Hv), electrical resistivity (p), and coefficient of thermal expansion (α) of $Ni_{1.04}Si_{1.98}$–$FeSi_2$ solid solutions at 20 °C. Sidorenko et al. (1968). With kind permission of Springer Nature

perature brittleness. Further, the low high temperature strength need to be improved. Attempts to achieve this goal pivoted around various additives, such as metallic components, solid solution forming elements, ceramic carbides or nitrides and last but not least addition of other silicides. In this section the effect of some important metallic elements are discussed among them Nb, Al and B.

Mechanical alloying was used to produce the desired aggregates. The powders of the desired composition were milled either in air or in flowing Ar for 200 h. The mechanical alloyed powders were placed in a graphite mold and were pre-pressed at room temperature. This was followed by pulse discharge sintering under a pressure of 55 MPa at 1673 K. Micrographs of $MoSi_2$ and $MoSi_2$-X (X = Al, B, Nb) are shown in Fig. 12.4 (sintering in air or Ar). X-ray diffraction patterns confirmed the compositions of the materials, which are: monolithic $MoSi_2$, $MoSi_2$-3 at.% Al, $MoSi_2$-5 at.% B and $MoSi_2$-5 at.% Nb (which are abbreviated as $MoSi_2$, 3Al, 5B and 5Nb). Vickers hardness is shown for the various materials tested in Fig. 12.5 and the fracture stress is seen in Fig. 12.6, respectively. The SEM fractographs of the tensile specimens are shown in Fig. 12.7. Both $MoSi_2$ specimens, from the commercial powder and the monolithic fractured transgranularly. On the other hand all the alloyed $MoSi_2$-X fractured in a mixed mode, but predominantly the fracture was intergranular. Whether the grain size or void formation due to imperfect sintering or residual stress (inducing crack formation) were the reason for the difference in fracture strengths was not definitely established, but any of the above could have been the reason. The fracture stress at 1273 K and a SEM micrograph of a $MoSi_2$-3Al alloy fabricated by the MA-PDS process after tensile test at 1273 K are illustrated in Figs. 12.8 and 12.9 respectively. The fracture of the $MoSi_2$-3%Al looks like a ductile fracture. Fine Al_2O_3 particles are dispersed in the structure, and thus the high strength of this alloy might be attributed to these particles.

The exceptional high temperature oxidation resistance of $MoSi_2$, low density and high thermal conductivity make the reinforced $MoSi_2$ a potential material for use.

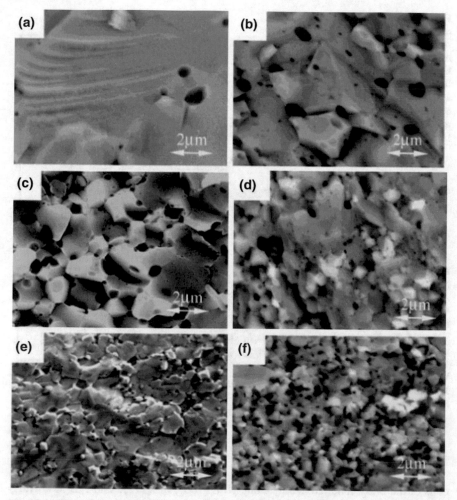

Fig. 12.4 SEM micrographs of MoSi$_2$-X (X = Al, B, or Nb) fabricated by mechanical alloying-pulsed discharge sintering (MA-PDS) process **a–d** are from pure MoSi$_2$, 3Al, 5B and 5 Nb milled in air; **e–f** are pure MoSi$_2$, 3Al, 5B and 5 Nb milled in Ar gas. Shan et al. (2002). With kind permission of The Japan Inst. Metals and Materials

12.3.2 Solid Solution Hardening (Softening)

Solid solution hardening and softening is observed in MoSi$_2$ depending on the alloying component. But before exploring these effects the addition of 1% Nb will be considered to provide an understanding on deformation on polycrystalline MoSi$_2$ tested by compression in the 25–1600 °C temperature range. One of the important characteristics by alloying with Nb is the fact that such an alloy can be plastically deformed by compression already at room temperature, while the unalloyed MoSi$_2$ fractures before plastic yielding at temperatures ≤900 °C. In Fig. 12.10 two unit

Fig. 12.5 Vickers hardness of MoSi₂ and MoSi₂-X (X = Al, B or Nb) alloy fabricated by MA-PDS process. The hardness of MoSi₂ sintered from commercial powders is also shown in the figure. Shan et al. (2002). With kind permission of The Japan Inst. Metals and Materials

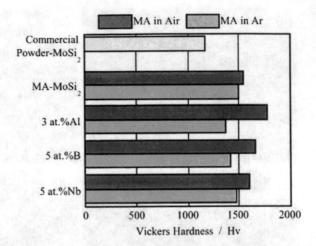

Fig. 12.6 Fracture stress of MoSi₂ and MoSi₂-X (X = Al, B or Nb) alloy fabricated by MA-PDS process. The fracture stress of MoSi₂ sintered from commercial powders is also shown in the figure. Shan et al. (2002). With kind permission of The Japan Inst. Metals and Materials

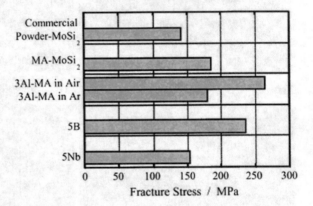

cells of the body centered MoSi₂ along with ⟨100⟩, 1/2⟨111⟩, and 1/2⟨331⟩ Burgers vectors are shown. Except for the hard orientated [001], single crystals can be deformed down to room temperature. However in polycrystalline MoSi₂ the lack of five independent slip systems has resulted in the brittlenes.

First principle calculations predicted that alloying with Al and Nb and other elements (Mg, V) may improve low temperature ductility by lowering the BDTT to ≤25 °C which was born out by experiments. Significant improvement was observed in the high temperature strength of MoSi₂ in particular of the high temperature yield strength.

In the case of Nb, addition to MoSi₂ has a dual effect, <600 °C solid solution softening occurs and at >900 °C hardening sets in. Unalloyed and alloyed MoSi₂ with Nb (Mo₀.₉₇, Nb₀.₀₃)Si₂ were arc melted in argon. Compression tests were performed in air by Instron at an initial strain rate of ~1 × 10⁻⁴ s⁻¹ from 25 to 1600 °C. True stress-true strain curves at the temperatures indicated deformed at 1% in compression tests are shown in Fig. 12.11. The 0.2% offset yield strength for the unalloyed and

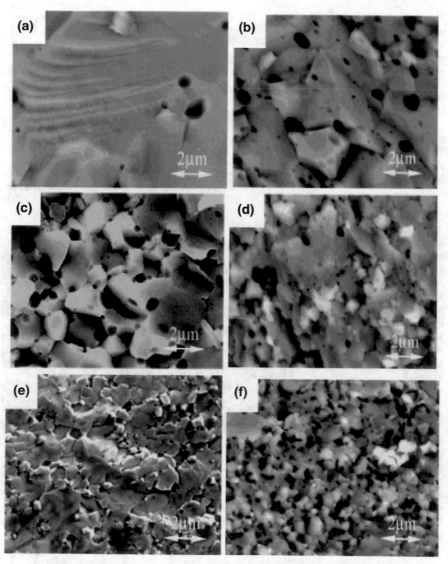

Fig. 12.7 SEM micrographs fracture surface after tensile test of MoSi$_2$ and MoSi$_2$-X (X = Al, B, Nb) alloy fabricated by MA-PDS process. **a** MoSi$_2$ made from commercial powders, **b** 3Al milled in air, **c–f** are those of 5B, pure MoSi$_2$, 3Al and 5 Nb milled in Ar gas. Shan et al. (2002). With kind permission of The Japan Inst. Metals and Materials

1% Nb alloyed MoSi$_2$ is seen in Fig. 12.12. The unalloyed MoSi$_2$ did not yield before brittle fracture below 900 °C as seen in Fig. 12.12. The yield strengths in the range 900–1600 °C of the unalloyed MoSi$_2$ polycrystals are 276 MPa and 14 MPa compared with 1045 MPa in single crystals. In the alloyed MoSi$_2$ at the temperature

Fig. 12.8 Fracture stress values of MoSi$_2$ and MoSi$_2$-X (X = Al or B) alloy fabricated by MA-PDS process. The tensile test is carried out at 1273 K. The fracture stress of MoSi$_2$ sintered from commercial powders is also shown in the figure. Shan et al. (2002). With kind permission of The Japan Inst. Metals and Materials

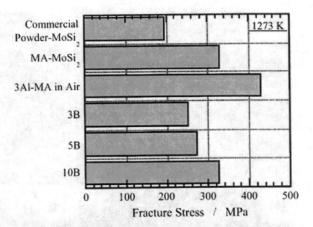

Fig. 12.9 SEM micrograph of fracture surface after tensile test of MoS$_2$-3 at.% Al alloy fabricated by MA-PDS process with powders milled in air. The tensile test is carried out at 1273 K. Shan et al. (2002). With kind permission of The Japan Inst. Metals and Materials

Fig. 12.10 Two unit cells of the body-centered tetragonal C11$_b$ structure of MoSi$_2$ and its important Burgers vectors. Sharif et al. (2003). With kind permission of Elsevier

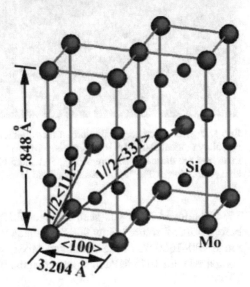

Fig. 12.11 True stress-true strain curves for 1 at.% Nb containing samples at various temperatures. Arrows indicates sample failure during compressive deformation testing. Sharif et al. (2003). With kind permission of Elsevier

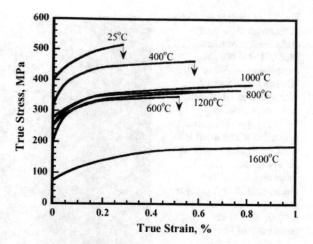

Fig. 12.12 Effects of alloying with 1 at.% Nb on the compressive yield strength of MoSi$_2$ in the 25–1600 °C temperature range. Polycrystalline MoSi$_2$ exhibited fracture before yield below 900 °C. Sharif et al. (2003). With kind permission of Elsevier

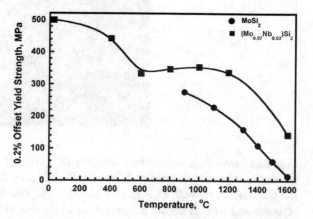

range 25–1600 °C the yield stress varies from 500 MPa to 1423 PMa, respectively as seen in Fig. 12.12.

The transition of BDTT was reduced from 900 °C to room temperature or even below it as a consequence of alloying with 1% Nb. Overall improvement of the yield strength at room temperature and 1600 °C (by an order of magnitude) occurred consequently. Note the small increase in the 0.2% offset yield strength in the temperature range 600–1200 °C. The transition of BDTT was reduced from 900 °C to room temperature or even below it as a consequence of alloying with 1% Nb.

The dislocation structure of the alloyed MoSi$_2$ is shown in Fig. 12.13 for 400 °C. The structure represents the temperature <600 °C where softening is observed consisting mainly of 1/2⟨111⟩ dislocations with some ⟨100⟩ dislocations. The reason that the polycrystalline MoSi$_2$ could not be deformed at low temperature by compression is related to the dislocation being mainly of ⟨100⟩ type. Thus it is possible to inferred that Nb might promote 1/2⟨111⟩ slip in MoSi$_2$. The insert in

Fig. 12.13 Bright field TEM micrograph showing the dislocation substructure in a MoSi$_2$-1 at.% Nb alloy compressed ~0.5% at 400 °C. The labels 1 and 2 correspond to ⟨100⟩, and 3 and 4 correspond to 1/2⟨111⟩ type dislocations, respectively. The insert is the weak beam image showing the dissociated 1/2⟨111⟩ dislocations. The line direction of the 1/2⟨111⟩ is generally ⟨110⟩. Sharif et al. (2003). With kind permission of Elsevier

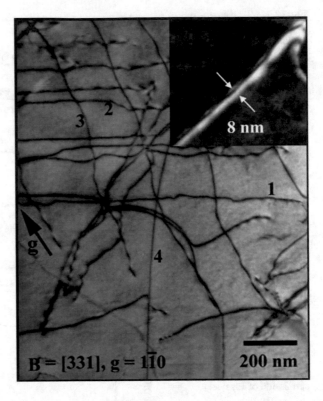

Fig. 12.13 reveals in weak beam imaging of 1/2⟨111⟩ dislocations that the partial dislocations spacing is increased, namely the stacking fault energy is lowered. Note the increase in strength in Fig. 12.12 in the temperature range of ~800–1100 °C, representing what is known the anomalous increase in yield strength with increasing temperature. Clearly no solution softening occurs in this region. The dislocation structure from a sample deformed at 800 °C is seen in Fig. 12.14. Of the several interpretations for the anomalous yield, one relates the observation as a first approximation to an "averaging" effect of slip on more than one type of slip systems, i.e., {110}⟨111⟩ and {011}⟨100⟩, in the polycrystalline MoSi$_2$. These dislocations showed a stronger tendency, as compared to samples deformed at 400 °C, to be straight and mostly along the 60° orientation, ⟨110⟩.

The magnitude of anomalous increase with temperature is reduced by Nb alloying compared to the unalloyed MoSi$_2$. The dislocation structure of specimens deformed at 1200 °C is seen in Fig. 12.15. At temperature >1600 °C the alloyed MoSi$_2$ is harder than the unalloyed and the dislocations are predominantly ⟨100⟩ in the form of dipoles and loops. The dislocations with ⟨110⟩ Burgers vectors are observed as reaction products of ⟨100⟩ and ⟨010⟩ dislocations. Following deformation at 1600 °C the dislocation substructure consists mainly of ⟨100⟩ and some 1/2⟨111⟩ as illustrated in Fig. 12.16. A comparison of the yield strength in single crystals MoSi$_2$ at the orientations indicated with polycrystalline-1 at.% Nb is made in Fig. 12.17.

Fig. 12.14 Bright field TEM micrograph showing the dislocation substructure in a MoSi$_2$-1 at.% Nb alloy compressed ~1% at 800 °C. The dislocations labeled x and y are both $1/2\langle 111\rangle$ type with line directions 60° from screw. $\mathbf{B} = [331]$, $\mathbf{g} = \bar{1}03$. Sharif et al. (2003). With kind permission of Elsevier

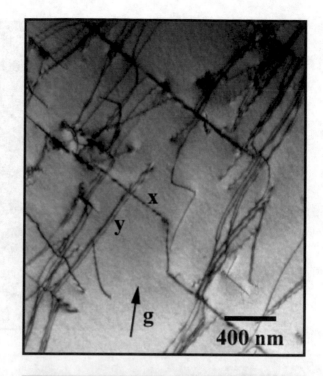

Fig. 12.15 Bright field TEM micrograph from a MoSi$_2$-1 at.% Nb alloy compressed ~1% at 1200 °C. All dislocations in this micrograph have a $\langle 100\rangle$ Burgers vector. Reactions between [100] and [010] produce [110] dislocations (e.g. marked with an arrow), and dipole loops that may have been 'pinched-off' from glide dislocations are also observed. Sharif et al. (2003). With kind permission of Elsevier

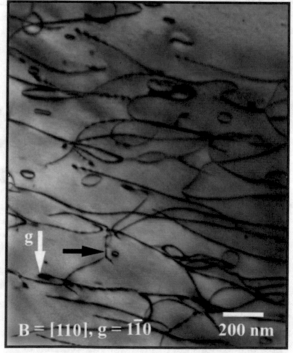

Fig. 12.16 Bright field TEM micrograph showing the dislocation substructure in a MoSi$_2$-1 at.% Nb alloy compressed at 1600 °C. An array of ⟨100⟩ dislocations forming a low angle boundary is marked with an arrow. **B** = [100] and **g** = 013. Sharif et al. (2003). With kind permission of Elsevier

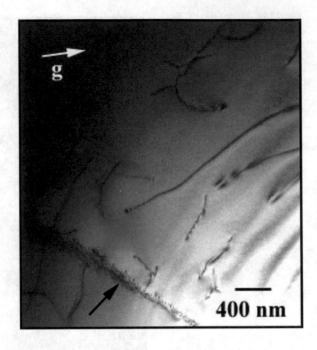

Fig. 12.17 Yield strengths for single crystals with [$\bar{1}$10] and [0 15 1] orientations from Ito et al. (1995) and the 0.2% offset yield strength of polycrystalline 1 at.% Nb-alloyed MoSi$_2$. Sharif et al. (2003). With kind permission of Elsevier

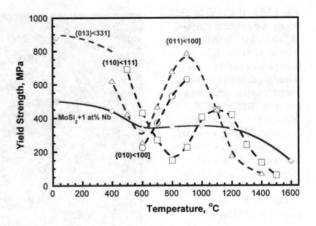

Unalloyed MoSi$_2$ has a very low yield strength at temperatures >1200 °C. It is believed that the reason for this is the relative easy glide of dislocations at high temperatures and climb is the rate controlling mechanism (well-defined dislocation cell structure). In Nb containing alloy deformation at 1200 °C results in dislocations with predominantly ⟨100⟩ and occasionally 1/2⟨111⟩ Burgers vectors-as mentioned earlier-and elongated loops formed from pinched off dipoles (Fig. 12.15). Sessile ⟨110⟩ dislocations are formed from the reaction of

$$\langle 100 \rangle + \langle 010 \rangle = \langle 110 \rangle$$

The absence of cell structure in the dislocation structure and the predominantly presence of tangles are responsible for the glide resistance in the alloy. Slip of $\langle 111 \rangle$ is not favored as it is at lower temperatures, where $\langle 100 \rangle$ slip is favored (Fig. 12.15).

Substitutional solute hardening in metals is usually expressed as

$$\Delta\tau = \frac{G\varepsilon^{3/2}x^{1/2}}{\alpha} \tag{12.1}$$

G is the shear modulus, ε is the Fleischer parameter, x is the concentration of the alloying element and α is a material constant (for most metals is in the order of 100s). The relation is to a first approximation the interaction of solutes with pure dilatational strain field only with edge dislocations and not with screw. Hence hardening by substitutional solutes is gradual, i.e. strength increases gradually with increasing solute concentration. The value of ε may be calculated from

$$\varepsilon = \left| \varepsilon'_G - \beta\varepsilon_a \right| \tag{12.2}$$

ε_a is the change in lattice parameter by concentration and given as

$$\varepsilon_a = (1/a)(da/dx) \tag{12.3}$$

The shear modulus is given by

$$\varepsilon'_G = \varepsilon_G/(1 + 0.5\varepsilon_G) \tag{12.4}$$

where ε_G is given by

$$\varepsilon_G = (1/G)(dG/dx) \tag{12.5}$$

$\beta = 3$ and 15 for edge and screw dislocations, respectively.

Hardening rates in the presence of defect complexes (in ionic crystals) create non-spherical strain fields. Such non-spherical strain fields have a stronger interaction with both edge and screw dislocations than spherical strain fields (considered in the classical hardening models). In MoSi$_2$ alloyed with Nb non-spherical strain fields may arise from: (a) electron deficiency (due to valence difference between Mo and Nb, i.e., 5 and 6 respectively) compensated by Mo vacancy creation, and (b) enhancing the clustering of Nb atoms to produce a local Nb-rich stoichiometry (i.e., (Nb, Mo)$_5$Si$_3$). Both Mo$_5$Si$_3$ and Nb$_5$Si$_3$ have a tetragonal D$_{8m}$ structure (space group I4/mcm). In the case of (b) the solute clustering leads to tetragonal distortion contrary to spherical elastic strain field around the Nb atoms in the lattice. Mitchell and Heuer (1977) have suggested for the dislocation-strain field interaction the relation given as:

$$\Delta \tau = \left(\frac{1.25 \Omega_d}{4\pi \Omega^{1/3} \lambda^2} \right)^{3/2} \left(\frac{\beta d}{b} \right)^{11/2} \left(\frac{M^3 \alpha^{3/2}}{\gamma} \right) \Delta \varepsilon^{3/2} G x^{1/2} \qquad (12.6)$$

$\Delta \tau$ is the increase in shear stress with increasing solute concentration, x, Ω_d is the defect complex volume, λ is the distance of the tetragonal distortion from the glide plane, $\beta = 2/3$ and 1 for screw and edge dislocations, respectively, d is the separation between the aliovalent cation and the charge compensating defect, $M = 1/(1 - \nu)$ for isotropic materials, $\alpha = (C_{11} - C_{12})/2C_{44}$ is the anisotropy factor, $\gamma = \Gamma/Gb^2$ where Γ is the line tension and b is the Burgers vector and $\Delta \varepsilon$ is the tetragonality of the defect complex, i.e., the deviation from spherical strain field.

It is believed that the high strengthening rate in MoS_2 is consistent with hardening due to tetragonality of the defect complex at the high temperature. At temperatures <1300 °C for unalloyed $MoSi_2$ glide dominates the deformation mechanism, climb dominates at temperatures >1500 °C (sub boundaries are seen in Fig. 12.16 a sign of climb) and in the temperature range ~1300–1500 °C a combination of glide and climb contribute to deformation as inferred from dislocation substructures and the stress exponents. Alloying with Nb (or Re) induces significant solution hardening as a result of complete change of the dislocation substructure which is consistent with glide controlled behavior as seen in Fig. 12.15 showing dislocation tangles and dipole loops contrary to unalloyed $MoSi_2$ where cell structure is obtained signifying climb controlled mechanism.

It can be concluded that in polycrystalline $MoSi_2$ Nb has a significant effect on yield strength in compression. At low temperature solution softening occurs leading to deformation even at room temperature. It is a result of reducing the stacking fault energy by increasing the ribbon width for $1/2\langle 111 \rangle$ dislocations and a lower barrier for kink nucleation on partial dislocations. At high temperatures of 1200–1600 °C the strengthening rate is consistent with glide controlled regime. Nb alloying leads to a significant reduction of the diffusion rates thereby suppressing climb contribution below 1600 °C. Tetragonality of te strain fields around the defects is associated with the hardening rate. At the intermediate temperature range of 600–1200 °C the yield strength of $(Mo_{0.97}, Nb_{0.03})Si_2$ remained almost independent of temperature due to the opposing trends of the critical resolved shear stress of the active slip systems, $\{011\}\langle 100 \rangle$ and $\{110\}\langle 111 \rangle$.

Alloying generally improves the mechanical properties as seen for the 0.2% yield stress and the tensile stress by the addition of 1% Nb, Re, ternary $(Mo, Re)Si_2$, Mo(Al, Si)$_2$, $(Mo, Nb)Si_2$ and quaternary $(Mo, Re)(Al, Si)_2$ and $(Mo, Nb)(Al, Si)_2$. Hardness and 0.2% offset yield stress are compared with pure $MoSi_2$ in Figs. 12.18 and 12.19. Surprisingly the microhardness values—except to the 2.5% Re containing alloy—are lower than that of the pure $MoSi_2$. The hardness values are 899 Hv, 729 Hv, 729 Hv and 104 Hv for the $MoSi_2$, 1 at.% Nb, 2 at.% Al, and 2.5% Re containing samples, respectively. Thus, except of the 2.5% Re alloy all had a lower microhardness than pure $MoSi_2$. Regarding the compressive yield stress at room temperature and at 1600 °C the following can be noted:

Pure $MoSi_2$ and the 2.5% Re alloy could not be deformed plastically below 900 and 1200 °C. Under these conditions, i.e., below these temperatures the samples

Fig. 12.18 Vickers hardness values for polycrystalline samples of all compositions under investigation. Hardness of polycrystalline Al containing samples is also compared to the values obtained on monocrystalline Al containing samples on (100) and (001) surfaces of the crystal. Sharif et al. (2001). With kind permission of Elsevier

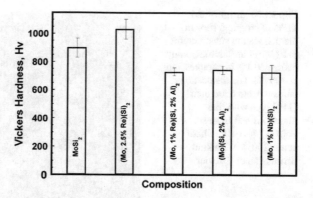

Fig. 12.19 Effects of alloying on the room temperature and high temperature (1600 °C) strength of MoSi$_2$. Sharif et al. (2001). With kind permission of Elsevier

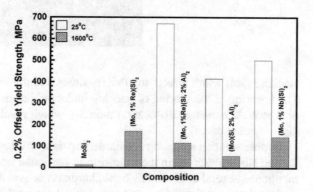

undergo brittle fracture. The BDTT of the MoSi$_2$ 2.5% Re alloy by compression increased by ~300 °C accompanied by an increase of the yield stress from 14 MPa to 170 MPa at 1600 °C. The most effective alloying element among those investigated in increasing the strength at 1600 °C is Re at a 2.5% level. As mentioned earlier at low temperatures < 600 °C softening is observed. The dislocation structure in 1 at.% Nb and also in 2 at.% Al exhibit mainly 1/2⟨111] slip with some ⟨100] slip. An example for 1% Nb is illustrated in Fig. 12.20 deformed at 800 °C. One of the suggestions for the Al softening is related to the lowering of the CRSS for slip on the {110}1/2⟨111] system. It was also observed that the addition of Al or Nb (also V and Ta) change the crystal structure from BCC tetragonal C11$_b$ to hexagonal C40 with a hexagonal atomic arrangement on the {110} planes, which occurs when their solubility limit is exceeded. Note that before the transformation, the MoSi$_2$ C11$_b$ crystal structure (space group I4/mmm) is characterized by the sequential stacking of the {110} atomic planes as ABAB. It is also known that 1/2⟨111] dislocations dissociated into two colinear 1/4⟨111] partials with stacking fault on {110} between them. The stacking fault has the sequence of ABCAB, which is the stacking sequence of C40 and closely related to that of the tetragonal C11$_b$. The alloying elements mentioned stabilize the C40 structure with a probable lowering of the stacking fault energy which also means increasing the width of the faulted region. The increase in

Fig. 12.20 Bright-field
TEM micrograph showing
the dislocation substructures
in a (Mo, 1 at.% Nb)Si$_2$ alloy
deformed 1% in compression
at 800 °C. The dislocations
labeled a and b are both
⟨111] type with line
directions 60° from screw: **b**
= [331], **g** = $\bar{1}$03. Sharif
et al. (2001). With kind
permission of Elsevier

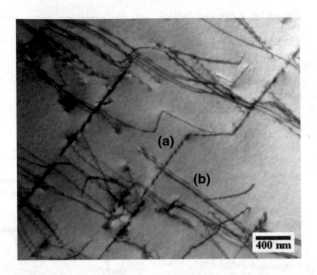

stacking fault width affects the Peierls stress and the mobility of the dislocations.
The lowering of the spacing of stacking faults on the other hand either by 1 at.% Nb
or 2 at.% Al enhances dislocation mobility which results in decreased flow stress,
i.e., softening occurs.

Further solution softening and hardening data on alloying of MoSi$_2$ in addition to
Re and Nb is presented in the following in particular consideration to hardness. As
mentioned several times in this book, hardness is associated with yield and tensile
stresses, but easier to perform and cost saving because the less complicated machines
are sufficient for the test. The components include alloys containing V, Cr, Zr, Nb,
Ta, W and Re and Mo as a constituent of the MoSi$_2$. Microhardness measurements
are compared in Fig. 12.21 for pure MoSi$_2$ and Mo$_{0.7}$M$_{0.3}$Si$_2$ where M stands for W
and Re.

The pure MoSi$_2$ is also shown in Fig. 12.21b for comparison. The error bars of
the measurements are seen for each curve. The microhardness of MoSi$_2$ decreases
with increasing temperature. The variation of the hardness with temperature can be
described by three stages: (a) the rapid decrease in hardness in the range 293–900 K;
(b) a gradual microhardness change in the temperature range 900–1100 K; and (c) a
second rapid decrease in hardness at the range 1100–1573 K. In the first stage many
slip traces and some microcracks were observed around the indentation. DBTT in
pure MoSi$_2$ is ~1173 K. The microhardness variation with temperature of the alloyed
MoSi$_2$ with Re and W follows the same pattern as the unalloyed MoSi$_2$ as seen in
Fig. 12.21b but at a higher hardness level. Both W and Re are completely soluble in
MoSi$_2$, thus we are observing what is known as solid solution hardening. The effect
of the other alloying elements are shown in Fig. 12.22.

Note that in Fig. 12.22a–e filled and open circles represent the alloys
(Mo$_{0.985}$M$_{0.015}$)Si$_2$ and (Mo$_{0.97}$M$_{0.03}$)Si$_2$, respectively, whereas in Fig. 12.22f these
symbols indicate Mo(Si$_{0.9925}$Al$_{0.0075}$)$_2$ and Mo(Si$_{0.985}$Al$_{0.015}$), respectively. The

Fig. 12.21 Temperature dependence of microhardness for **a** pure MoSi$_2$ and **b** (Mo$_{0.7}$M$_{0.3}$)Si$_2$ (M = W, Re). Harada et al. (1998). With kind permission of Elsevier

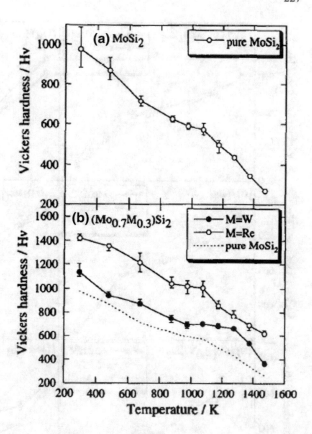

shape of these curves can also be divided into three stages, thus following the same pattern as that of MoSi$_2$ seen earlier in Fig. 12.21a. There was no hardening above about 1173 K and the segments of the graphs roughly are at the same level as the unalloyed MoSi$_2$. Since these elements are soluble to some extent in MoSi$_2$ we can speak of solid solution strengthening.

The compositional dependence of the microhardness in W and Re alloyed MoSi$_2$ is seen in Fig. 12.23. Re and W alloying does not change the tetragonal C11$_b$ structure.

The hardness increase with the concentration of W and Re, but a small peak occurs at ~0.9 followed by a decreases of hardness to a level of the pure alloying constituent. The peak is more pronounced in the case of Re. The observations in Fig. 12.23 substantiates the solid solution hardening concept on the Mo-rich side of alloying with Mo(Re$_{(1-x)}$, W$_x$)Si$_2$ or (Mo$_{1-x}$Re$_x$)Si$_2$.

The compositional dependence of the microhardness at 293 K of MoSi$_2$ alloyed with Nb, Ta, Cr, Zr, V or Al is illustrated in Fig. 12.24. The alloying elements have limited solubility and consequently a second phase is observed in the alloys where the added element exceeds the solid solubility. The phases are C40-type or C49. These phases are indicated at the bottom of the hardness curves of Fig. 12.24 as would be present in the constitution diagram. In the two phase region the hardness increases

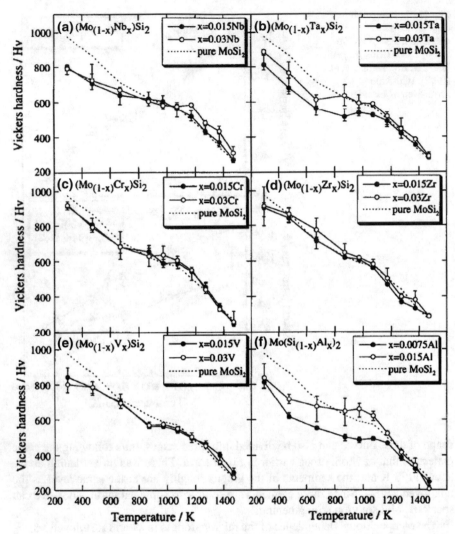

Fig. 12.22 Temperature dependence of microhardness for **a** (Mo, Nb)Si$_2$, **b** (Mo. Ta)Si$_2$, **c** (Mo, Cr, Si)$_2$, **d** (Mo, Zr)Si2, **e** (Mo, V)Si$_2$ and **f** Mo(Si, Al)$_2$. Harada et al. (1998). With kind permission of Elsevier

except in the case of Al where a decrease in hardness is seen in Fig. 12.24f. There are only limited data for V (Fig. 12.24e), but since it is in the same periodic group as Nb and Ta, it is expected to show similar behavior. In the one phase region (tetragonal C11$_b$) there is a decrease in hardness. It is thus clear that solution softening occurs in the C11$_b$ region regardless of the alloying elements except of W and Re (Fig. 12.23). In the case of W and Re additions hardening occurs.

The alloying effect on the microhardness by the various elements are assembled in Fig. 12.25. The hardness of the two alloys (Mo$_{0.985}$M$_{0.015}$)Si$_2$ and (Mo$_{0.97}$M$_{0.03}$)Si$_2$,

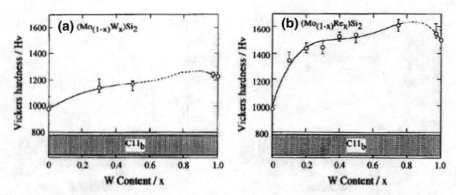

Fig. 12.23 Compositional dependence of microhardness for MoSi$_2$ alloyed with **a** W and **b** Re. Harada et al. (1998). With kind permission of Elsevier

respectively are compared at 293 K with the hardness of pure MSi$_2$ where M stands for the elements mentioned earlier. ΔH_1 and ΔH_2 of the figures are defined in Eq. 12.7 where the compounds in the parenthesis in the equation represent their hardness.

$$\Delta H_1 = H_V\left(Mo_{(1-x)}M_xSi_2\right) - H_V(MoSi_2)$$
$$\Delta H_2 = H_V(Msi_2) - H_V(MoSi_2) \tag{12.7}$$

Note the linear relation in Fig. 12.25a, b regardless if structural change has occurred. Recall that the structure did not change on alloying with Re and W (Fig. 12.23) which remained tetragone C11$_b$, while the other elements promote C40 and C49 type compounds formation.

Some elements induce solid solution softening at room temperature as indicated above. The effect was explained by increased dislocation mobility due to scavenging impurities from the matrix on one hand while the possibility of reducing the Peierls stress and nucleating screw dislocations by alloying were also suggested to explain softening.

12.4 Alloying of FeSi$_2$

12.4.1 Metallic Element Addition

The potential application of iron silicides is for thermoelectric conversion devices by the utilization of the temperature difference effectively in the 230–630 °C range. Therefore most of the publications are related to their use as thermoelectric material, the improvement of their thermoelectric efficiency, thermal stability at high temperature and chemical stability. The low temperature phase β-FeSi$_2$ is orthorhombic and is an intrinsic semiconductor. Extrinsic thermoelectric FeSi$_2$ can be prepared

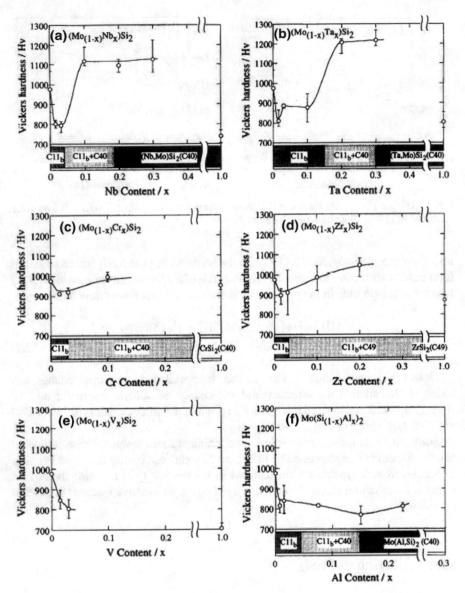

Fig. 12.24 Compositional dependence of microhardness for MoSi$_2$ alloyed with **a** Nb, **b** Ta, **c** Cr, **d** Zr, **e** V and **f** Al. Harada et al. (1998). With kind permission of Elsevier

by doping; thus for example by doping with Al or Mn p-type thermoelectricity is obtained, while Co addition results in an n-type thermoelectric material.

Powder mixture of Fe, Si and Co were mechanically alloyed for 120 h under Ar atmosphere to yield Fe$_{0.98}$Co$_{0.02}$Si$_2$ to obtain an n type iron silicide. A SEM micrograph of the Fe$_{0.98}$Co$_{0.02}$Si$_2$ powder is shown Fig. 12.26. The as-milled powders

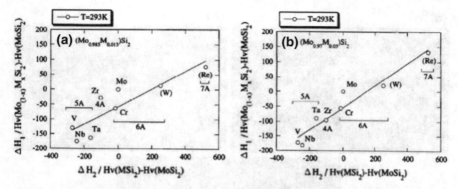

Fig. 12.25 Correlation in the microhardness at 293 K between alloyed MoSi$_2$ and disilicides. Harada et al. (1998). With kind permission of Elsevier

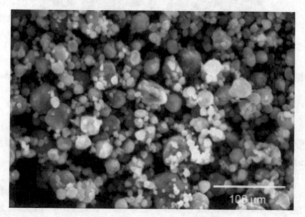

Fig. 12.26 SEM micrograph of Fe$_{0.98}$Co$_{0.02}$Si$_2$ powders, mechanically alloyed for 120 h. Ur et al. (2002). With kind permission of Springer Nature

were consolidated by hot pressing (after degassing in vacuum for 2 h at 400 °C) at 1000 °C at a stress of 35–60 MPa for 2 h and at 1100 °C for 4 h. After final fabrication the typical chemical composition of the mechanically alloyed powder was 67.12 at.% Si, 32.23 at.% Fe and 0.65 at.% Co. Transformation to β-FeSi$_2$ occurs in stages with a peritectoid reaction also involved. Therefore the stages are

Primary transformation   α-Fe$_2$Si$_5$ + ε-FeSi → β-FeSi$_2$ below 865 °C

Secondary transformation   Si + ε-FeSi → β-FeSi$_2$ (12.8)

The progressive phase transformation of the Fe$_{0.98}$Co$_{0.02}$Si$_2$ powder at various thermal conditions is illustrated in Fig. 12.27. The white area is identified as ε-FeSi and that of the matrix is a mixture of α and β phases of FeSi$_2$. As seen in Fig. 12.27 the ε-FeSi diminishes with the annealing time indicating that the reaction is progressive.

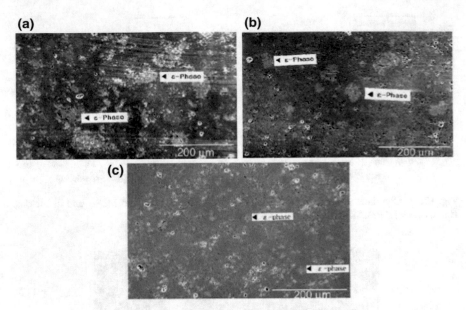

Fig. 12.27 SEM micrograph of n-type $Fe_{0.98}Co_{0.02}Si_2$ compacts indicating progressive phase transformation during annealing: **a** as-VHPed, **b** 830 °C/12 h and **c** 830 °C/48 h. Ur et al. (2002). With kind permission of Springer Nature. VHP stands for vacuum hot pressing

That the reaction did not go to completion and the phases mentioned above were still present at annealing of 830 °C for 24 h can be seen in the TEM micrographs shown in Fig. 12.28.

In Fig. 12.28a, A denotes β-FeSi$_2$, B marks Fe$_2$Si$_5$ and C represent ε phase. The structure is fine grained of about 2 μm grain size. Stacking faults are also seen in this figure VHP stands for vacuum hot pressing. Full transformation to β-FeSi$_2$ is very slow due to the peritectoid reaction involving diffusion and may take very long time to full transformation, perhaps even of hundreds of hours. Application of VHP at 60 MPa at 1100 °C can reduce transformation to ~24 h. The microhardness in as VHPed Fe$_{0.98}$Co$_{0.02}$Si$_2$ increased with time and saturated at around 12 h as seen in Fig. 12.29.

No additional information on other mechanical properties beyond hardness is available on the effect of addition to FeSi$_2$, although as a general comment it is indicated that the mechanical properties improved during isothermal annealing under VHP. This is probably due to the interest in the thermoelectric properties of iron silicide.

12.5   Alloying of TiSi$_2$

Titanium silicides are of interest as structural material for coatings, turbine airfoils, burning chamber parts and missile nozzles among other uses for high temperature

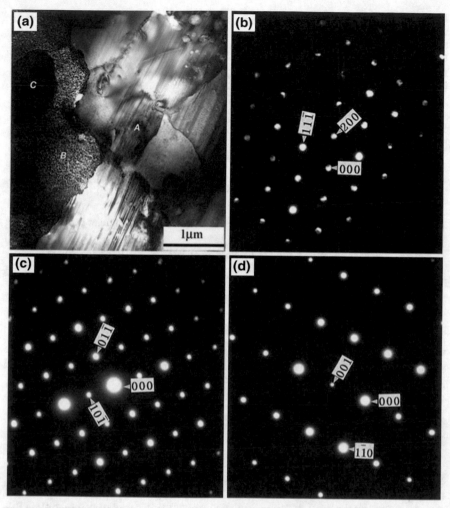

Fig. 12.28 TEM micrograph of Fe$_{0.98}$Co$_{0.02}$Si$_2$ compacts annealed at 830 °C for 24 h: **a** bright field image, **b** [011] SAD pattern from β-FeSi$_2$, **c** [111] SAD pattern from α-Fe$_2$Si$_5$ and **d** [011] SAD pattern from ε-FeSi. Ur et al. (2002). With kind permission of Springer Nature

applications because of their excellent oxidation and hot gas corrosion resistance. A vast amount of publications on TiSi$_2$ film dealt mainly with its use for VLSI applications with understandable emphasis on resistivity, chemical stability, transformation kinetics from the C49 to the C54 modification and many additional subjects related to its electronic applications. Although stress, hardness and creep strength are important features of such films, almost no research appears in the literature or very limited studies are sporadically found on mechanical properties. In Table 12.2 the strength properties of TiSi$_2$ alloyed with SiC are shown in sample 1. Sample 2 refers to TiSi$_2$–SiC–Ti$_3$SiC$_2$. Compare the fracture toughness value of sample 1 in Table 12.2

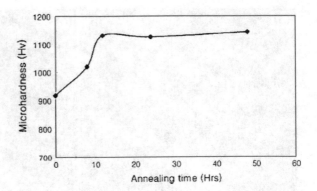

Fig. 12.29 Micro-hardness variation in $Fe_{0.98}Co_{0.02}Si_2$ compacts during annealing. Ur et al. (2002). With kind permission of Springer Nature

Table 12.2 Mechanical properties of specimens. Qin et al. (2006). With kind permission of the Japan Institute of Metals and Materials

Specimen	Vickers microhardness, Hv (GPa)	Fracture toughness K_{IC} (MPa $m^{1/2}$)	Bending strength σ_b (MPa)
Sample 1	11.5 ± 0.3	3.3 ± 0.2	400 ± 50
Sample 2	12.1 ± 0.2	5.4 ± 0.3	700 ± 60

with that of 2.1 MPa $m^{1/2}$ of pure $TiSi_2$ (Rosenkranz and Frommeyer 1992). The samples were in situ fabricated by spark plasma sintering resulting only in $TiSi_2$ and SiC according to the relation $3Si + TiC \rightarrow TiSi_2 + SiC$.

12.6   Alloying of Ti_5Si_3

In the light of the importance of the Ti based materials for high temperature structural applications, the lack of sufficient information on the mechanical properties of alloyed $TiSi_2$ and the availability of such data for Ti_5Si_3, it was felt that including this specific material in this chapter will add an additional dimension for understanding the developing tendency of using Ti based materials. Add to this, the very high melting point of 2138 °C and the low density of 4.32 g/cm^3 of Ti_5Si_3 we might expect an ideal material for high temperature applications. Alloyed Ti_5Si_3 with Nb and Cr are considered in this section.

Alloys were prepared by arc melting followed by homogenization anneal under vacuum at 1000 °C for 30 h. The nominal compositions of the alloys investigated is shown in Table 12.3 indicating that the alloys contain Nb and Cr (X in the alloys).

As seen in Table 12.4 stoichiometric and off-stoichiometric alloys were investigated. The needle-like microstructure of Ti_5Si_3 is compared with the Cr alloyed microstructures in Fig. 12.30. The needles in the unalloyed Ti_5Si_3 are straight forming a net structure within the grains without crossing the grain boundaries. In the Cr alloyed structure the amount of needles decrease with increasing the Cr con-

Table 12.3 Nominal compositions of the alloys. Zhang and Wu (1998). With kind permission of Elsevier

Alloy label	Composition (at.%)
Stoichiometric alloys (Ti$_{62.5-X}$M$_X$Si$_{37.5}$)	
Ti$_5$Si$_3$	Ti–37.5Si
2Nb	Ti–2Nb–37.5Si
5Nb	Ti–5Nb–37.5Si
10Nb	Tiv10Nb–37.5Si
20Nb	Ti–20Nb–37.5Si
1Cr	Ti–lCr–37.5Si
2Cr	Ti–2Cr–37.5Si
5Cr	Ti–5Cr–37.5Si
Off-stoichiometric alloys (Ti$_{62.5}$M$_X$Si$_{37.5-X}$)	
Off-5Nb	Ti–5Nb–32.5Si
Off-l0Nb	Ti–10Nb–27.5Si
Off-2Cr	Ti–2Cr–35.5Si

centration, disappearing form the grains and segregating in the grain boundaries as determined by energy dispersive X-ray spectrum (EDX) and electron probe micro-analyzer (EPMA) and shown in Fig. 12.31. The segregation of Cr into grain boundaries is also seen in Table 12.4. The primary phase in the alloys is Ti$_5$Si$_3$ which dissolves some amounts of the alloying elements as can be in Tables 12.5 and 12.6 for the stoichiometric and off-stoichiometric alloys. The relation between Vickers hardness and the composition of the stoichiometric Ti$_5$Si$_3$ alloys is seen in Fig. 12.32. As known a relation between the flow stress and hardness exists and expressed by the relation

$$H = C \cdot Y \qquad (12.9)$$

where H is the hardness, Y is the flow stress and C is a factor. It can be seen that at low alloying the hardness, i.e. the flow stress increases but beyond a certain amount it decreases. The increasing part of the curve is a result of solution strengthening, and the flow stress decline is attributed at least in the case of Cr to its inhomogeneous distribution.

Three Nb containing phases, labeled "A", "B" and "C" are seen in Table 12.6 with different morphologies. The figures of these phases are seen in Fig. 12.33b, c.

Figure 12.33a relates to Cr alloying. The temperature dependence of the hardness of the alloyed Ti$_5$Si$_3$ is compared with unalloyed Ti$_5$Si$_3$ in Fig. 12.34. It seems that the effect of composition in both the Nb and Cr alloyed T$_5$Si$_3$ is more pronounced than at the higher temperatures, i.e., above 400 °C. The hardness lines are relatively close to each other above this temperature, and also the change in the hardness values with temperature is less pronounced. Further tests to evaluate the mechanical properties of the alloyed Ti$_5$Si$_2$ were performed by compression at a strain rate of 1×10^{-4} s^{-1}. The fracture toughness of the off-stoichiometric alloys is illustrated in Fig. 12.35

Table 12.4 Phase constitutions and compositions of the stoichiometric alloys. Zhang and Wu (1998). With kind permission of Elsevier

Alloy label	Phase constitution (X-ray diffraction)	Composition (at.%)		
		Average	Matrix	Grain
Ti_5Si_3	Ti_5Si_3, monolithic	Ti, 61.2	65.8	60.6
		Si, 38.8	34.2	39.4
2Nb	Ti_5Si_3, monolithic	Ti, 61.8		
		Nb, 1.8		
		Si, 36.4		
5Nb	Ti_5Si_3, monolithic	Ti, 58.5		
		Nb, 4.4		
		Si, 37.1		
10Nb	Ti_5Si_3, monolithic	Ti, 53.4		
		Nb, 9.7		
		Si, 36.9		
20Nb	Ti_5Si_3, monolithic	Ti, 44.4	42.5	50.5
		Nb, 18.8	21.0	8.6
		Si, 36.6	36.5	40.9
1Cr	Ti_5Si_3, monolithic	Ti, 62.9	62.0	52.7
		Cr, 1.4	0.9	4.6
		Si, 35.6	37.1	42.7
2Cr	Ti_5Si_3, monolithic	Ti, 60.9	61.2	54.3
		Cr, 2.0	1.3	8.1
		Si, 37.1	37.5	37.6
5Cr	Ti_5Si_3, monolithic	Ti, 58.4	59.8	50.7
		Cr, 4.5	2.7	11.9
		Si, 37.2	37.5	37.4

for the 5Nb, 10Nb and 2Cr off-stoichiometry alloys as a function of temperature. One can observe that the strength decreases for 10Nb continually with temperature, while for the 5Nb alloy a sharp decrease occurs above 400 °C. Up to that point an increase in strength is seen. In the case of Cr additive, the strength decrease occurs above ~700 °C and up to this point a slight fracture strength improvement occurs, but the overall strength is below that of the off-stoichiometric Nb alloyed Ti_5Si_3. The fracture surfaces of a binary Ti_5Si_3 with the Nb alloyed specimen are compared in Fig. 12.36. The fracture surface indicates a cleavage fracture with river pattern and secondary cracks are observed. This figure contradicts some researchers point of view that the fracture behavior in the binary Ti_5Si_3 is transgranular for large grains and intergranular for small grains. The fracture surfaces of the off-stoichiometric alloys after compression are shown in Fig. 12.37.

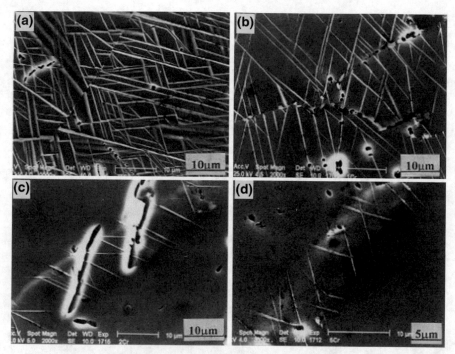

Fig. 12.30 SEM micrographs of the binary Ti$_5$Si$_3$ and Cr alloyed stoichiometric alloys after etching. **a** Ti$_5$Si$_3$, **b** alloy 1Cr, **c** alloy 2Cr and **d** alloy 5Cr. Zhang and Wu (1998). With kind permission of Elsevier

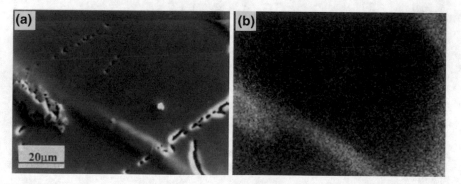

Fig. 12.31 Cr mapping in alloy 5Cr showing the grain boundary is Cr enriched. **a** Morphology and **b** Cr mapping. Zhang and Wu (1998). With kind permission of Elsevier

Table 12.5 Composition of the primary Ti$_5$Si$_3$ phase in the off-stoichiometric alloys. Zhang and Wu (1998). With kind permission of Elsevier

Alloy label	Off-2Cr	Off-5Nb	Off-10Nb
Composition (at.%)	Ti, 60.8	T$_i$, 55.4	Ti, 52.7
	Cr, 0.6	Nb, 5.4	Nb, 8.2
	Si, 38.7	Si, 39.1	Si, 39.1

Table 12.6 Composition of the eutectic area in the off-stoichiometric alloys. Zhang and Wu (1998). With kind permission of Elsevier

Alloy label	Composition (at.%)			
	Average	"A" (Ti₅Si₃)	"B" (Nb-rich phase)	"C" (Ti-rich phase)
Off-5Nb	Ti, 78.0	69.2	87.6	95.0
	Nb, 6.0	5.2	10.4	3.7
	Si, 16,0	25.5	2.0	1.3
Off-10Nb	Ti, 67.9	64.9	80.3	87.6
	Nb, 12.2	9.5	16.2	8.7
	Si, 19.9	25.6	3.5	3.8
Off-2Cr	Ti, 79.0	–	–	–
	Cr, 19.1			
	Si, 1.8			

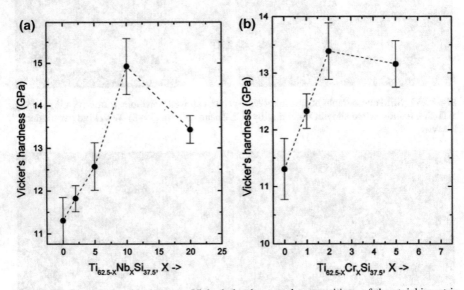

Fig. 12.32 The relationship between Vicker's hardness and compositions of the stoichiometric Ti₅Si₃ alloys. **a** Nb alloying and **b** Cr alloying. Zhang and Wu (1998). With kind permission of Elsevier

The fracture surfaces of the primary Ti₅Si₃ are smooth with many river patterns on them as seen in Fig. 12.37a, b, while the eutectic area (Fig. 12.37b, c) indicates fine intergranular fracture. Fracture occurs in both regions, the primary Ti₅Si₃ phase and in the boundary between the primary Ti₅Si₃ phase and the eutectic area.

It is possible to express the alloy formed by the addition of Nb or Cr to Ti₅Si₃ as the formula (Ti, M)₅Si₃ (M stands for the additive). Stoichiometric Ti₅Si₃ can dissolve large amounts of Nb or Cr and the alloy is composed of monolithic Ti₅Si₃ with high

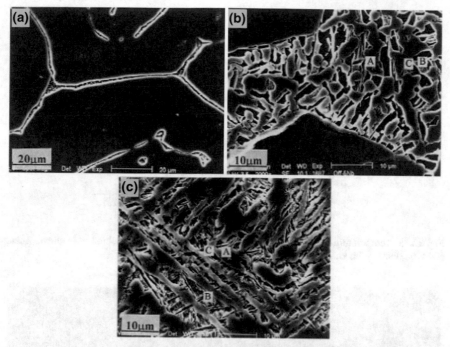

Fig. 12.33 SEM micrographs of the eutectic area in the off-stoichiometric alloys. **a** Alloy off-2Cr, **b** alloy off-5Nb and **c** alloy off-10Nb. Zhang and Wu (1998). With kind permission of Elsevier

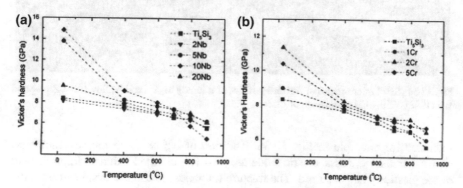

Fig. 12.34 Microhardness versus temperature. **a** Nb alloying and **b** Cr alloying. Zhang and Wu (1998). With kind permission of Elsevier

content of these elements even up to 21 and 11.9 at.%, respectively while the off-stoichiometric alloys containing Nb and Cr dissolved in Ti$_5$Si$_3$ as the primary phase. The designation of the formula as indicated [(Ti, M)$_5$Si$_3$] means that the alloying elements replace part of Ti in the alloy.

Usually increase in strength is associated with a decrease of elongation at all temperatures. Ti$_5$Si$_3$ is a brittle compound and cracks develop in the material as

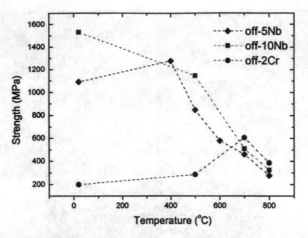

Fig. 12.35 Temperature dependence of the fracture strength for off-stoichiometric alloys. Zhang and Wu (1998). With kind permission of Elsevier

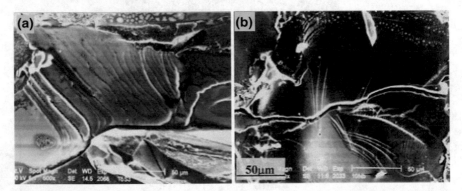

Fig. 12.36 SEM micrographs of fracture surfaces of **a** binary Ti$_5$Si$_3$ and **b** alloy 10Nb. Zhang and Wu (1998). With kind permission of Elsevier

indicated for example in Fig. 12.36. Addition of Nb or Cr to the stoichiometric alloys increase hardness, elastic modulus, yet—it is claimed—that no improvement of the plasticity has occurred. The fracture toughness is usually evaluated from the crack length and the hardness according to the relation of

$$K_{Ic} = \zeta \left(\frac{E}{H}\right)^{1/2} \left(\frac{P}{c^{2/3}}\right) \qquad (12.9)$$

where $\xi = 0.016 \pm 0.004$. E is the elastic modulus, H is the hardness, P is the indentation load and c is the crack length measured from the center of the indentation. The cracks generally are observed in the vicinity or around the hardness indentation. This relation was used for the stoichiometric fracture toughness evaluation. The

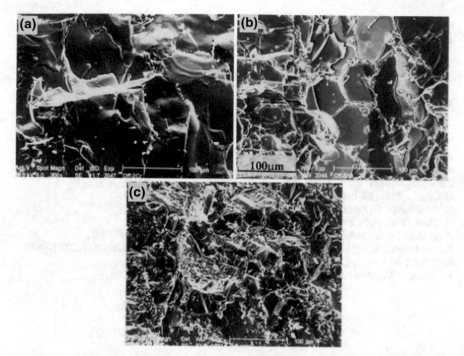

Fig. 12.37 SEM micrographs of fracture surfaces of off-stoichiometric alloys. **a** Alloy off-2Cr, **b** alloy off-5Nb and **c** alloy off-10Nb. Zhang and Wu (1998). With kind permission of Elsevier

off-stoichiometric alloy is less brittle than the stoichiometric alloy, which can be attributed to its hypereutectic microstructure where its volume depend on the amount of Nb or Cr added. Despite this improvement of the off-stoichiometric alloy its elevated temperature strength is not sufficiently high to ensure very high temperature application. One should note that the above elements form substitutional rather than interstitial solid solution. Thus whatever improvement in strength of Ti_5Si_3 by Nb or Cr is observed, it is a result of substitutional solid solution hardening.

Summary

- Alloying to enhance strength properties with elements, such as Al, Nb, Cr and Re
- Hardening rates in the presence of defect complexes create non-spherical strain fields
- Non-spherical strain fields are more effective strengthener; they interact with both edge and screw dislocations
- High temperature strenghtening ($MoSi_2$) is due to tetragonality of the defect complex
- $MoSi_2$ have exceptional oxidation resistance
- Nb and Re induce solid solution strengthening in $MoSi_2$ by changing the dislocation substructure

- Solid solution softening occurs by Al additions to $MoSi_2$
- Ti_5Si_3 is an ideal material for high temperature applications.

References

Y. Harada, Y. Murata, M. Morinaga, Intermetallics **6**, 529 (1998)
K. Ito, H. Inui, Y. Shirai, M. Yamaguchi, Philos. Mag. A **72**, 1075 (1995)
A.A. Sharif, A. Misra, T.E. Mitchell, Mater. Sci. Eng. A **358**, 279 (2003)
T.E. Mitchell, A.H. Heuer, Mater. Sci. Eng. **28**, 81 (1977)
C. Qin, L. Wang, W. Jiang, S. Bai, L. Chen, Mater. Trans. **47**, 845 (2006)
R. Rosenkranz, G. Frommeyer, Mater. Sci. Eng. A **152**, 288 (1992)
A. Shan, W. Fang, H. Hashimoto, Y.-H. Park, Mater. Trans. **43**, 5 (2002)
A.A. Sharif, A. Misrab, J.J. Petrovicb, T.E. Mitchell, Intermetallics **9**, 869 (2001)
F.A. Sidorenko, L.A. Miroshnikov, P.V. Gel'd, Sov. Powder Metall. Met. Ceram. **7**, 292 (1968)
S.-C. Ur, I.-H. Kim, J.-I. Lee, Met. Mater. Inter. **8**, 169 (2002)
L. Zhang, J. Wu, Acta Mater. **46**, 3535 (1998)

Chapter 13
Grain Size Effect on Mechanical Properties

Abstract Hardness tests is a simple, cost saving method to estimate mechanical properties of material, requiring commonly available equipment and therefore an attractive method. Relationships between hardness and other mechanical properties exist, the well known one is that of Tabor for the tensile stress evaluation. This chapter emphasizes the grain size effect on the mechanical properties. Small grain size improves static properties such as tensile and yield stresses while large grains enhance creep resistance. A main factor in concentrating on $MoSi_2$ and Ti_5Si_3 is the availability of data but also due to their excellent high temperature properties. The Hall–Petch relation between hardness and grain size is discussed in this chapter.

13.1 Introduction

It is a common knowledge that grain size affects the mechanical properties of polycrystalline material. Small grains strengthen some mechanical properties, such as tension, whereas large grains improve creep resistance. In the following we shall attempt to present tensile properties for the silicides available in the literature as affected by the grain size, and also the effect of large grain size on the creep properties will be discussed.

13.2 MoSi$_2$ Static Properties

The hardness of monolithic $MoSi_2$ is included in Fig. 13.1 expressing the Hall–Petch relation between hardness and size$^{-1/2}$. Here grain size is expressed by the symbol L which is used in this section also to be true to the original presentation. As known one of the relations between hardness and ultimate tensile stress is that of Tabor given as

$$\sigma_u = \frac{H}{2.9}[1 - (m - 2)]\left[\frac{12.5(m - 2)}{1 - (m - 2)}\right]^{(m-2)} \tag{13.1}$$

© Springer Nature Switzerland AG 2019
J. Pelleg, *Mechanical Properties of Silicon Based Compounds: Silicides*,
Engineering Materials, https://doi.org/10.1007/978-3-030-22598-8_13

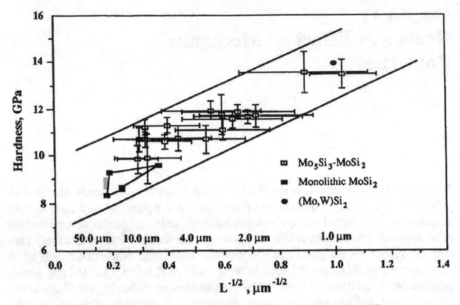

Fig. 13.1 Plot showing the Hall–Petch relationship between hardness and $L^{-1/2}$, where L refers to the nominal width of the $MoSi_2$ phase in the lamelar eutectic alloys indicated and grain size of the monolithic $MoSi_2$. Note the room temperature hardness values of the monolithic $MoSi_2$ (and of $(Mo,W)Si_2$) fall within the linear band of the eutectic material. Mason and Van Aken (1993). With kind permission of Elsevier

where H is the diamond pyramid hardness, m is Meyer's hardness coefficient. It has been shown that $n = (m - 2)$, where n is the strain hardening coefficient. Replacing $(m - 2)$ by n one can rewrite (13.1) as

$$\sigma_u = \frac{H}{2.9}(1 - n)\left(\frac{12.5n}{1 - n}\right)^n \tag{13.2}$$

Hardness tests are an attractive method to estimate other mechanical properties such as tensile stress or yield stress because it is simpler and cost effective. Hardness values for several grain sizes for monolithic $MoSi_2$ are presented in Table 13.1. Two surprising observations can be made in this list: (a) the single crystal monolithic $MoSi_2$, has a higher value than the specified polycrystalline crystals, (b) the grain size variation almost don't influence the hardness values of the specified $MoSi_2$ structures except the alloy in Ref. 15 and (c) smaller grain size does not seem to increase the hardness (namely tensile strength) as commonly known to occur (see for example alloys of Refs. 2 and 7). At this stage there is no point of speculating on this seemingly not expected results, since several explanations can be given. The most direct approach is to check the appropriate references. Careful control of the purity level is essential for obtaining reproducible results. An example of the impurities effect can be observed in the fracture toughness data as a function of the grain

Table 13.1 Hardness and grain size values of monolithic MoSi₂. Mason and Van Aken (1993). With kind permission of Elsevier

Hardness (GPa)	Grain size	Refs.
11.80	Not specified	Alman et al. (1992)
8.38	30 μm	Gibala et al. (1992)
10.78 11.84	Not specified	Tiwari and Herman (1992)
8.58	18 μm	Maloy et al. (1991)
10.13	Not specified	Castro et al. (1992)
10.37		
12.42	Not specified	Keiffer et al. (1952)
8.70	18 μm	Wade (1992)
9.60	7 ± 4 μm	Schwartz et al. (1992)
13.9[a]	1 μm	
9.25	28 μm	Bhattacharya and Petrovic (1991)
9.86	Single crystal	Boldt et al. (1992)

[a](Mo,W)Si₂

size shown in Fig. 13.2. Here it is assumed that the lower line—thus lower fracture toughness—is a reflection of the impurities present in the grain boundaries inducing a glassy phase reducing the strength. The upper line in the figure corresponds to a purer starting powder. The results indicate that careful control of purity of the source powder is essential to obtain high strength and toughness. During the fabrication (compaction, sintering, annealing) the material is exposed to elevated temperature and therefore grain growth might occur. An example of the effect of temperature on

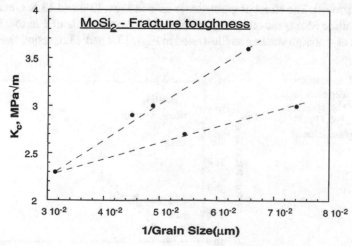

Fig. 13.2 Fracture toughness variation with grain size in MoSi₂. Sadananda et al. (1999). With kind permission of Elsevier

Fig. 13.3 Kinetics of grain growth in MoSi$_2$ during hot pressing. Sadananda et al. (1999). With kind permission of Elsevier

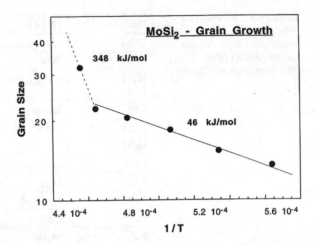

grain growth in MoSi$_2$ of different grain sizes is seen in Fig. 13.3. It is clearly creep conditions, and the figure substantiates the expectations that large grain size is more resistant to creep.

13.3   MoSi$_2$ Creep (Time Dependent) Properties

As already seen in Fig. 13.3 creep performs better at large grain size which is more resistant to creep damage. The extreme case is clearly the single crystal where the large grains coalesce into one single unit. In the absence of reinforcing agents control of grain size can modify a material and with increased grain size its creep behavior can be improved. The effect of grain size is seen in Figs. 13.4 and 13.5. Generally in polycrystalline MoSi$_2$ the creep exponent n varies from a value of 1 at low stresses, to a value of 4 at high stresses as illustrated in Figs. 13.4 and 13.5, respectively. Also

Fig. 13.4 Load-history effects on creep rates for larger grain size specimen. Sadananda et al. (1999). With kind permission of Elsevier

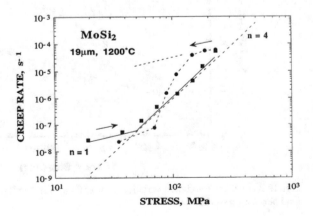

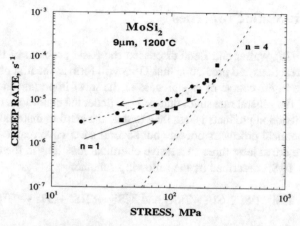

Fig. 13.5 Load-history effects on creep rates for smaller grain specimen. Sadananda et al. (1999). With kind permission of Elsevier

observe that the creep rate is much influenced by the grain size variation in particular in the region where n = 1. A decrease of 3–4 orders of magnitude is seen with only a small increase in grain size.

Specifically Fig. 13.6 indicates the effect of grain size on creep rate whatever the strain exponent is. Thus MoSi₂ having 25 μm has the lowest creep rate at the same stress as the creep rate of a 14 μm MoSi₂ which deforms at a larger rate by about three orders of magnitude. Thus, polycrystalline MoSi₂, possesses elevated temperature creep behavior which is highly sensitive to grain size. Tailoring the grain size of MoSi₂ can modify creep resistance even without additives.

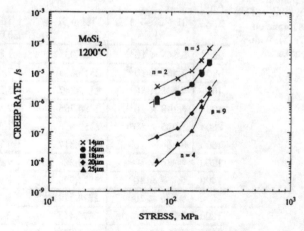

Fig. 13.6 Creep rate of polycrystalline MoSi₂, as a function of grain size. Petrovic (1995). With kind permission of Elsevier

13.4 Ti$_5$Si$_3$ Static Properties

Already in (1976) Walser and Bené discussed the basic problem of the first phase nucleation in reactions and predicting that TiSi$_2$ will form as the first phase. The free energy change in the course of any process is the most important thermodynamic parameter and the general rule states that $\Delta G < 0$ under the conditions where the free energy of the reaction products is less than that of the starting materials. In silicides as in other chemical processes not one but several successive reactions occur. For example, there are at least three successive chemical reactions in the conversion of Ti and Si into Ti$_5$Si$_3$ described by the following scheme:

$$Ti + Si \rightarrow TiSi, \ TiSi + Si \rightarrow TiSi_2 \ and \ TiSi_2 + TiSi + 3Ti \rightarrow Ti_5Si_3. \quad (13.3)$$

As indicated the effectiveness of the chemical reactions is determined by the value of ΔG, and beside other important factors such as the kinetics, the more ΔG negative is the more favorable the reaction is. As known the Gibbs free energy is given by

$$\Delta G = \Delta H - T\Delta S \quad (13.4)$$

The expression combines enthalpy and entropy. The values of ΔG_T^0 are useful to calculate the equilibrium constant of the respective reaction and is given as

$$\Delta G_T^0 = -RT \ln K_{eq} \quad (13.5)$$

The equilibrium constants given in Table 13.2 for the reactions in the Ti–Si system indicated above are applicable also for the system obtained by the self propagating

Table 13.2 Equilibrium constants of the reactions of titanium silicides formation. Strogova et al. (2015). With kind permission of Dr. Strogova

K_p			
T, K	Ti + Si $\rightarrow$ TiSi	TiSi + Si $\rightarrow$ TiSi$_2$	TiSi$_2$ + TiSi + 3Ti $\rightarrow$ Ti$_5$Si$_3$
298	$3,645 \times 10^{23}$	2524,03	$1,879 \times 10^{51}$
300	$2,561 \times 10^{23}$	2513,225	$7,874 \times 10^{50}$
400	$4,821 \times 10^{17}$	2197,347	$5,503 \times 10^{36}$
600	$6,148 \times 10^{11}$	2086,286	$2,585 \times 10^{22}$
800	$5,925 \times 10^{8}$	2157,326	$1,331 \times 10^{15}$
1000	$8,204 \times 10^{6}$	2280,317	$4,758 \times 10^{10}$
1200	$4,386 \times 10^{5}$	2422,388	$4,611 \times 10^{7}$
1400	$5,137 \times 10^{5}$	2571,939	$2,993 \times 10^{5}$
1600	$9,876 \times 10^{3}$	2724,112	$6,444 \times 10^{3}$
1800	2655,202	2876,415	310,698
2000	905,496	3027,958	26,457
2200	367,948	3178,142	3,419

high temperature sintering technique. The change in the free energy, ΔG_T^0 with temperature for the three reactions shown in Eq. 13.3 is illustrated in Fig. 13.7. In Fig. 13.7 formation of Ti₅Si₃ according to the last reaction in Eq. 13.3 is favorable at least at temperatures <~1600 °C having a large negative free energy. Note the preferential formation of Ti₅Si₃ compared with the prediction of Walser and Bené.

Mechanically amorphized Ti–Si powders were dynamically densified to obtain the compacts which were then recrystallization-annealed in the range 800–1200 °C for 1–12 h to obtain a single phase Ti₅Si₃. The grain size was fine as determined by TEM and XRD analysis.

The grain size changed from 50 nm of the 1 h 800 °C annealed specimen to 160 nm of the heat treated samples at 1220 °C for 3 h (Fig. 13.8a). The grain size of the specimens almost did not grow and remained stable at 115–125 nm when annealed at 1000 °C for up to 12 h Fig. 13.8b). Figure 13.8 shows the Ti₅Si₃ crystallite variation with time at constant temperature and with temperature at constant time. The Vickers microhardness variation with grain size is presented in Fig. 13.9 in two ways. In Fig. 13.9a the hardness variation with the square root of the grain size, namely the inverse Hall–Petch square root relation is shown while in Fig. 13.9b the Vickers hardness values are plotted versus the grain size directly. Note that in these figures the plots are at three impact velocities, those of 300, 500 and 700 m/s. At least 10 values of hardness measurements were obtained. The standard deviations of the data were in the range of 2–5% of the average values. The fracture toughness data in the grain size range 6–50 nm are accumulated from various references in Table 13.3. Note that the fracture toughness values increase with decreasing grain size. This is expected, since the smaller the grain size the static strength properties

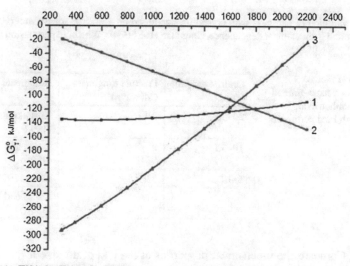

1– TiSi, 2 –TiSi₂, 3 – Ti₅Si₃

Fig. 13.7 Temperature dependence ΔG_T^0 for the silicide formation reactions. Strogova et al. (2015). With kind permission of Dr. Strogova

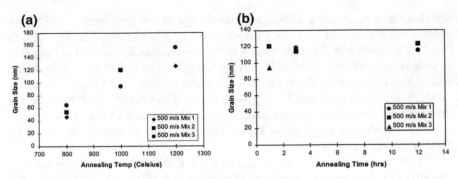

Fig. 13.8 Plot showing crystallite size variation with **a** annealing temperature at constant time (1 h), and with **b** annealing time at constant temperature (1000 °C). Counihan et al. (1999). With kind permission of Elsevier. m/s stands for the impact velocity

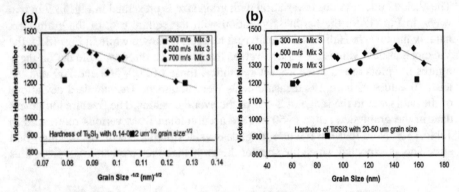

Fig. 13.9 Variation of hardness with crystallite size based on **a** inverse-square root (Hall–Petch) dependence and **b** direct linear dependence. Counihan et al. (1999). With kind permission of Elsevier

Table 13.3 Fracture toughness of coarse-grained Ti_5Si_3. Counihan et al. (1999). With kind permission of Elsevier	Grain size (μm)	Fracture toughness (MPa√m)	References
	20–50	2.1	Rosenkranz et al. (1992)
	10–30	2	Ruess and Vehoff (1990)
	6–10	3.2	Mitra (1998)
	6	5	Ruess and Vehoff (1990)

increase. Compare the mechanical properties at the μm grain size level which are listed in Table 13.4 with the nanograin size information shown earlier (Fig. 13.9 and Table 13.3). Table 13.4 includes alloyed Ti_5Si_3 also. The various mechanical properties of Ti_5Si_3 in the grain size range of 5–10 μm are listed in Table 13.4. The

Table 13.4 Mechanical Properties of Ti$_5$Si$_3$, Ti$_5$Si$_3$-20 Vol Pct TiC composite, and Ti$_5$Si$_3$-8 Wt Pct Al alloy. Mitra (1998). With kind permission of Springer Nature

Material/property	Ti$_5$Si$_3$	Ti$_5$Si$_3$-20 Vol Pct TiC	Ti$_5$Si$_3$-8 Wt Pct Al
Grain size	5–10 μm	5–10 μm	5–10 μm
Hardness	12.7 ± 0.5 GPa	14.4 ± 0.2 GPa	13.5 ± 0.5 GPa
Flexural strength	255 ± 5 MPa	–	–
Fracture toughness	3.2 ± 0.1 MPa $\sqrt{m}$	4.1 ± 0.1 MPa $\sqrt{m}$	4.15 ± 0.3 MPa $\sqrt{m}$
Compressive yield strength			
1100 °C, 10^{-3}/s	1235 MPa	–	220 MPa
1200 °C, 10^{-3}/s	550 MPa	–	–

Table 13.5 Hardness of Ti$_5$Si$_3$ having different grain sizes. Mitra et al. (2008). With kind permission of Springer Nature

Method of processing	Grain size (μm)	Hardness (GPa)	References
Powder metallurgy	1–2	17.1 ± 0.7	Thom et al. (1994)
Powder metallurgy	5–10	12.7	Mitra (1998)
Powder metallurgy	5–6	11.7 ± 0.8	Min et al. (1995)
Arc melting and annealing	20–30	11.3 ± 0.5	Zhang and Wu (1998)
Powder metallurgy	20–50	9.5 ± 0.3	Rosenkranz et al. (1992)

hardness variation with grain size is assembled in Table 13.5 obtained by various researchers. Note that the average hardness for Ti$_5$Si$_3$ could differ by about 10–20% of similar grain size. This is likely to be due to the presence of flaws like microcracks. Also the yield strength is grain size dependent as indicated in Table 13.6.

13.5 Creep (Time Dependent) Properties

High melting point materials such as Ti$_5$Si$_3$ are candidates for elevated temperature applications. Ti$_5$Si$_3$ in particular is of interest because of its high melting point of 2130 °C and low density of 4.3 g/cm^3 (Mitra 1998) and good oxidation resistance at and below 850 °C. However, most of the work is related to alloyed Ti$_5$Si$_3$, and almost no published research can be found in the literature.

Table 13.6 Yield strength obtained by compression testing in the Ti_5Si_3 with different grain sizes under different strain rates. Mitra et al. (2008). With kind permission of Springer Nature. Ti_5Si_3–TiC composite is included

Material	Grain size (μm)	Strain raters (s^{-1})	Yield strength (MPa)		References
			At 1100 °C	At 1200 °C	
Ti_5Si_3	5–10	10^{-3}	1115	550	Mitra (1998)
Ti_5Si_3	25	10^{-4}	826	609	Rosenkranz et al. (1992)
Ti_5Si_3-35 TiC composite	4	4×10^{-4}	800	400[a]	Li et al. (2002)

[a]Indicates data calculated by interpolation

The general observation regarding grain size effect on the static properties (strength, toughness, etc.) is the trend summarized by the Hall–Petch relation, namely there is an increase in the strength properties with decreasing grain size. This is to say, fine grain size strengthens the material. However creep rate increases with decreasing grain size. Thus, contrary to the Hall–Petch relation, a prerequisite for good creep resistance is a large grain size material. Plots showing the steady state creep rate at 1200 °C with the applied stress for Ti_5Si_3 of 5 and 20 μm grain sizes (also for two alloys) is shown in Fig. 13.10.

It can be observed that a specific stress, the Ti_5Si_3 specimens with the larger grain size have the higher resistance to creep. In other words, it is clear that the Ti_5Si_3 specimens with 20 μm grain size show a lower creep rate than the Ti_5Si_3 having 5 μm size grains. The power law relation of the creep rate has been presented in Eq. (6.8). It is inversely related to the grain size, d. Under the same conditions, namely the same stress, the same grain size and stress exponents, the smaller the

Fig. 13.10 Plots showing the variation of the logarithm of creep rate with the logarithm of stress. Mitra et al. (2008). With kind permission of Springer Nature. Ti_5Si_3-Al and Ti_5Si_3-TiB$_2$ composites are included

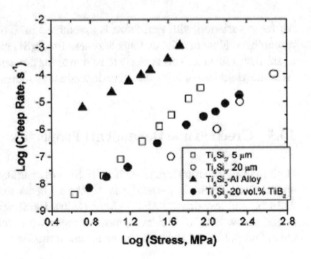

grain size is the larger is the creep rate. Thus it is understood why polycrystals with large grain size are essential for improved creep resistance.

As indicated earlier creep is a time dependent deformation. General description of creep was discussed in Chap. 6. All materials at low stress and high temperature show a Newtonian flow and the stress exponent is n = 1. With increasing stress, when n = 3–4 stage II creep sets in (the power-law creep). The transition from Newtonian to power-law creep depends on grain size and load history. Knowing the grain size exponent and the activation energy enables the identification of the creep rate controlling mechanism. Figure 13.10—as indicated—is a plot showing the effect of the grain size on the variation of the creep rate with stress.

Relation (6.8) expresses also the connection of creep rate with activation energy. The activation energy for creep in the 25 μm grain size Ti_5Si_3 in the temperature range 800–1200 °C is $350 \pm 20\ kJmol^{-1}$, while that of the 5 μm grain size is 620–640 $kJmol^{-1}$ as the applied stress is reduced from 26 to 3.5 MPa at 1200–1400 °C. Since creep is associated with diffusion it is likely that the higher activation energy >1200 °C is connected with lattice diffusion playing role in the creep deformation. At the lower temperatures where climb is more seldom, dislocation-dislocation interactions and thermally activated glide of jogs in screw dislocations are the more likely mechanisms responsible for the deformation.

Summary

- Grain size effect on Mechanical properties of $MoSi_2$ and Ti_5Si_3 (rather than $TiSi_2$)
- Static and time dependent properties (creep) are affected by grain size
- Large grains size material are more resistant to creep than small size grains
- At low temperatures dislocation-dislocation interaction while at high temperature climb control creep deformation
- At high temperatures where climb occurs diffusion plays an important role.

References

D.E. Alman, K.G. Shaw, N.S. Stoloff, K. Rajan, Mater. Sci. Eng. A **155**, 95 (1992)
A. Bhattacharya, J.J. Petrovic, J. Am. Ceram. Soc. **74**, 2700 (1991)
P.H. Boldt, J.D. Embury, G.C. Weatherly, Mat. Sci. Eng. A **155**, 251 (1992)
R.G. Castro, R.W. Smith, A.D. Rollett, P. Stanek, Scr. Met. **26**, 207 (1992)
P.J. Counihan, A. Crawford, N.N. Thadhan, Mater. Sci. Eng. A **267**, 26 (1999)
R. Gibala, A.K. Ghosh, D.C. Van Aken, D.J. Srolovitz, A. Basu, H. Chang, D.P. Mason, W. Yang, Mater. Sci. Eng. A **155**, 147 (1992)
R. Keiffer, E. Cerwenka, Z. Metal. **43**, 101 (1952)
J. Li, D. Jiang, S. Tan, J. Eur. Ceram. Soc. **22**, 551 (2002)
S. Maloy, A.H. Heur, J. Lewandoski, J. Petrovic, J. Am. Ceram. Soc. **74**, 2704 (1991)
D.P. Mason, D.C. Van Aken, Scripta Met. Mater. **28**, 185 (1993)
K.S. Min, A.J. Ardell, S.J. Eck, F.C. Chen, J. Mater. Sci. **30**, 5479 (1995)
R. Mitra, Met. Mater. Trans. A **29A**, 1629 (1998)

R. Mitra, N. Eswara Prasad, Y.R. Mahajan, Trans. Indian Inst. Met. **61**, 427(2008)

J.J. Petrovic, Mater. Sci. Eng. A **192/193**, 31 (1995)

R. Rosenkranz, G. Frommeyer, W. Smarsly, Mater. Sci. Eng. A **152**, 288 (1992)

S. Ruess, H. Vehoff, Scripta Metall. Mater. **24**, 1021–1026 (1990)

K. Sadananda, C.R. Feng, R. Mitra, S.C. Deevi, Mater. Sci. Eng. A **261**, 223 (1999)

R.B. Schwartz, S.R. Srinivasan, J.J. Petrovic, C.J. Maggiore, Mat. Sci. Eng. A **155**, 75 (1992)

A.S. Strogova, A.A. Kovalevskii, O.M. Komar, AASCIT J. Mater. **1**, 123 (2015)

A.J. Thom, M.K. Meyer, Y. Kim, M. Akinc, in *Processing and Fabrication of Advanced Materials III*, ed. by V.A. Ravi, T.S. Srivatsan, J.J. Moore et al. (TMS, Warrendale, PA, 1994), p. 413

R. Tiwari, H. Herman, Mater. Sci. Eng. A **155**, 95 (1992)

R.K. Wade, J. Am. Ceram. Soc. **75**, 1682 (1992)

R.M. Walser, R.W. Bené, Appl. Phys. Lett. **28**, 15 (1976)

L. Zhang, J. Wu, Acta Mater. **46**(10), 3535 (1998)

Chapter 14
Environmental Effect

Abstract Oxidation in $CoSi_2$, $NiSi_2$, $MoSi_2$, WSi_2 and $TiSi_2$ was evaluated by Rutherford backscattering spectra. Oxidation was carried out in dry and wet (steam) oxygen. A protective SiO_2 forms during the process but the silicide remains untact. Almost all disilicides posses excellent oxidation and corrosion resistance due to the stable SiO_2 formation. From kinetic measurements it is evident that SiO_2 formation is much larger in steam oxidation than in dry oxygen. Stress may be induced in the silicide during its formation due to the deposition parameters, thermal expansion mismatch and the presence of contaminants. $MoSi_2$ and WSi_2 coatings are used for protection against corrosion and oxidation at elevated temperatures. C addition to $MoSi_2$ improves hardness, fracture toughness and creep properties. SiC strengthened $TiSi_2$ composite increases hardness, fracture toughness and yield strength among other beneficial effects.

14.1 Introduction

The detrimental effect of certain environments have been known for many years. In this section the atmospheric effects is considered, which induces failures by corrosion, oxidation etc. Atmospheric effects, such as corrosion have been reported to account for more failure in terms of soundness degradation than any other types of damage. Composition of the alloy, the presence of minute surface fissures and surface discontinuities (invisible cracks to the naked eye) are all nascent features of environmental failure. The silicide types and their compositions considered in this book are the subject of the environmental effect exploration. Oxidation often forms a protecting atmosphere on the surface stabilizing it and thus reducing the rate of failure or even eliminating it.

14.2 CoSi₂-Oxidation

Oxidation was carried out between 750 and 1100 °C in a quartz tube furnace through which steam or dry oxygen flowed at atmospheric pressure. During oxidation under such conditions good SiO_2 layers could be formed on $CoSi_2$.

© Springer Nature Switzerland AG 2019
J. Pelleg, *Mechanical Properties of Silicon Based Compounds: Silicides*,
Engineering Materials, https://doi.org/10.1007/978-3-030-22598-8_14

Kinetic measurements show that SiO_2 formation on the silicides is much larger during steam oxidation than during oxidation in dry oxygen. Tracer experiments using radioactive silicon show that silicon is very mobile in the silicide during oxidation and that SiO_2 formation takes place at the silicide-SiO_2 interface, the oxidant diffusing through the growing SiO_z layer. In general silicides have excellent conductivity and therefore they are used as ohmic or rectifying contacts as interconnects. It is necessary in such application to cover the silicide with an insulating layer, such as SiO_2, to enable the use of overcrossing conductors in subsequent layers. Therefore the oxidation of a silicide, such as CoS_2 its stability during oxidation is of great interest. The structural stability, chemical stability and kinetics of the thermal oxidation of $CoSi_2$ were measured by using radioactive Si tracer.

Co was deposited onto $Si\langle 100\rangle$ substrate by electron gun evaporation in a vacuum of $\sim 5 \times 10^{-7}$ Torr. Subsequent silicide formation was done by annealing at a pressure of 1×10^{-6} Torr. The thermal oxidation was carried out between 750 and 1100 °C in steam or dry oxygen at a constant flow rate in atmospheric pressure. The structural stability, the chemical stability and the thickness of the SiO_2 formed, was investigated by Rutherford backscattering of 2 meV α particles. Figure 14.1 shows the Rutherford backscattering spectra before and after oxidation in steam at 900 °C. The $CoSi_2$ layer is stable, stays intact and an SiO_2 layer forms at the surface. It is chemically stable as indicated by the absence of metal oxide formation. The chemical stability is expressed as

$$\text{Percentage chemical stability} = \frac{A_{ms} - A_{mo}}{A_{ms}} \frac{100}{1}\% \qquad (14.1)$$

A_{ms} is the area under the metal signal before oxidation and A_{mo} the area under the surface position metal signal after oxidation (i.e. the metal oxide signal). Table 14.1 shows the percent chemical stability of silicides during oxidation. $CoSi_2$ is also listed in the Table at different oxidation temperatures. The structural stability in

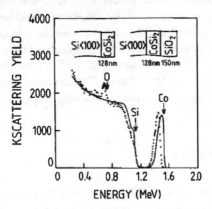

Fig. 14.1 Rutherford backscattering spectra of $CoSi_2$ silicide before (-) and after (···) thermal oxidation in steam at 900 °C. Strydom and Lombaard (1985). With kind permission of Elsevier

Table 14.1 Percent chemical stability of silicides during steam oxidation at different temperatures for 20 min. Strydom and Lombaard (1985). With kind permission of Elsevier

Temperature (°C)	Chemical stability (%) of the following silicides					
	CoSi$_2$	CrSi$_2$	NiSi$_2$	PtSi	TiSi$_2$	ZrSi$_2$
1100	100	97	100	100	87	0
1050	100	98	100	100	91	0
1000	100	99	100	100	90	0
950	100	97	100	100	94	0
900	100	99	100	100	89	0
850	100	98	100	100	90	0

steam oxidation is seen in Fig. 14.2 and in Table 14.2 dry oxidation is also listed. It is given by the relation as

$$\text{Percentage structural stability} = \frac{B}{A}\frac{100}{1}\%\tag{14.2}$$

A and B are the areas determined by integration of the Co metal signal before and after oxidation, respectively. The structural stabilities of silicides during oxidation, including that of CoSi$_2$ are listed in Table 14.2. The steam and dry oxidations were for a constant time of 20 min and 40 min, respectively. CoSi$_2$ exhibits excellent oxidation resistance and some ductility even in steam oxidation as expressed by structure stability indicated in Table 14.2. The chemical stability is 100% even at the high temperature of 1100 °C. The oxidation was evaluated in the temperature

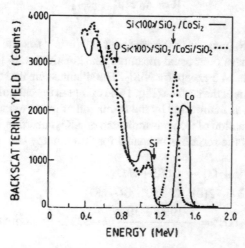

Fig. 14.2 Spectra showing steam oxidation of CoSi$_2$ on inert SiO$_2$ substrate. The spectra for CoSi$_2$, was measured at a target angle of 60°. Strydom and Lombaard (1985). With kind permission of Elsevier

Table 14.2 Percentage structural stability for silicides. Strydom and Lombaard (1985). With kind permission of Elsevier

Temperature (°C)	Structural stability (%) for the following silicides				
	$CoSi_2$	$CrSi_2$	$NiSi_2$	PtSi	$TiSi_2$
Oxidation in steam (20 min)					
1100	55	65	–	–	80
1050	62	72	–	–	90
1000	100	93	–	–	93
950	100	100	56	75	100
900	100	100	83	91	100
850	100	100	100	100	100
Oxidation in dry oxygen (40 min)					
1100	97	71	–	–	91
1050	100	78	–	–	98
1000	100	100	45	60	100
950	100	100	73	80	100
900	100	100	94	94	100

range 850–1100 °C. The rate of oxidation is presented in Fig. 14.3 as Arrhenius plots expressing the temperature dependence of the process involved.

Relation for the oxidation can be expressed as the Arrhenius relation is

$$R = R_0 \exp\left(-\frac{Q}{kT}\right) \tag{14.3}$$

In the relation R $(m^2 s^{-1})$ is the rate and R_0 $(m^2 s^{-1})$ is the rate constant and the other symbols have their usual meaning. The kinetic results for the silicides are summarized in Table 14.3 except for $NiSi_2$. The silicides remain intact and during the oxidation SiO_2 forms at the surface (Fig. 14.1) except in the case of Ti. In this case the chemical instability is manifested by the formation of metal oxide also (Fig. 14.19). In Fig. 14.2 the formation of $CoSi_2$ is indicated on SiO_2 substrate. Formation of SiO_2 can occur by any of the reactions indicated for the enthalpy change:

$$CoSi_2 + O_2 \rightarrow CoSi + SiO_2$$
$$3(-34) \quad 2(0) \quad 2(-50) \quad 3(-287)$$
$$\therefore \quad \Delta H° = \frac{-1962 + 102}{5} = -372 \, \text{kJ (g atom O)}^{-1} \tag{14.4}$$

The $CoSi_2$ layer stays intact during oxidation, the SiO_2 layer forms at the surface (Fig. 14.1). Tracer experiments using radioactive silicon tracer show that the silicon

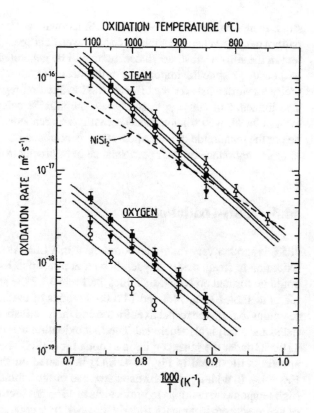

Fig. 14.3 Arrhenius plots showing the temperature dependence of the oxidation rates in steam and dry oxygen of PtSi (△), CoSi₂ (■), CrSi₂ (○) and TiSi₂ (▼) formed on Si⟨100⟩ substrates. Oxidation rates in steam are between 20 and 60 times higher. Strydom and Lombaard (1985). With kind permission of Elsevier

Table 14.3 Kinetics of silicide oxidation. Strydom and Lombaard (1985). With kind permission of Elsevier

Silicide	Oxidation in steam		Oxidation in dry oxygen	
	R_0 (m²s⁻¹)	Q (eV)	R_0 (m²s⁻¹)	Q (eV)
NiSi₂	2.1×10^{-13}	0.99 ± 0.2	–	–
TiSi₂	1.3×10^{-11}	1.4 ± 0.2	2.0×10^{-13}	1.3 ± 0.3
CrSi₂	1.6×10^{-11}	1.4 ± 0.2	1.0×10^{-13}	1.3 ± 0.3
CoSi₂	1.7×10^{-11}	1.4 ± 0.2	2.7×10^{-13}	1.3 ± 0.3
PtSi	2.0×10^{-11}	1.4 ± 0.2	2.3×10^{-13}	1.3 ± 0.3

is very mobile and its thickening at the silicide-SiO₂ interface takes place by the oxidant diffusing through the growing SiO₂ layer.

Very large amount of work appeared in the literature ever since the applicability of silicides,—among them CoSi₂—as an important component in the microelectronics was discovered, in particular in very large scale integration (VLSI). The mechanical characteristics related to the silicide formation (known as salicidation) is a critical parameter; stress is induced during the salicidation process which is accompanied with the generation of dislocations in the substrate silicon. This occurs when the local stress increases as a consequence of scaling down the dimensions and increasing the pattern density. When the stress exceeds the critical shear stress on an active slip system in the silicon substrate, dislocations will be generated as mentioned and defects will occur. Despite the importance of the stress (or other mechanical properties) virtually no work exists for the critical stress evaluation regardless of the silicide type. The influence of alloying to strengthen the silicide cannot be underestimated, but even in the alloyed silicides no systematic research work on the mechanical properties or the magnitude of the critical stress is published except of sporadic statements to their importance on their performance in microelectronics.

14.3 $NiSi_2$-Oxidation

$NiSi_2$ is another very important component in VLSI. As in the case of $CoSi_2$, during oxidation in steam or dry oxygen flowed at atmospheric pressure good SiO_2 layers could be formed on $NiSi_2$. In Tables 14.1 and 14.2 the stability at various temperatures, in Tables 14.3, 14.4 and 14.5 the kinetics of oxidation, the enthalpies for the reactions to form the respective silicide and their activation energies, respectively are listed and $NiSi_2$ is also included. Thermal oxidation was carried out between 750 and 1100 °C under the same condition as done for $CoSi_2$. The Rutherford backscattering spectra is illustrated in Fig. 14.4. SiO_2 is formed on the $NiSi_2$ surface as seen in Fig. 14.4. In addition to their extensive use in the salicidation, silicides are used as high temperature coatings to protect gas turbine hot section components from attack of aggressive environments. Clearly the protective character at elevated temperatures stems from the protective SiO_2 formation. Almost all disilicides posses excellent oxidation resistance due to the stable SiO_2 formation. The structural stability can be seen in Fig. 14.5, where the oxidation was performed in steam and the substrate is SiO_2.

Internal stress may develop in the silicide due to the deposition conditions. Annealing may induce substantial stress in the forming silicides. This is an important issue, since mechanical stress can alter the electrical properties. In addition stresses may develop as a consequence of: (a) deposition parameters, (b) difference in thermal expansion between the substrate (usually silicon) and the silicide, (c) presence of contaminants (or dopants, oxygen in the film is a contaminant), (d) structure and composition of the film (an example is shown on this aspect for $TaSi_2$). When the substrate is polycrystalline silicon and the film is a polycide, cracks can develop in the film and lifting of the film can occur due to the stress exceeding a critical value. An example for the stress variation in the silicide (as mentioned for $TaSi_2$) is illustrated in Fig. 14.6. Note the high stress at crystal structure change (transition); also note that the presence of grains relaxed the stress since it accommodates stress in the grain boundaries. Contaminants in a silicide changes its properties. Oxygen is a contaminant and its effect on the $TaSi_2$ film properties can be seen in Fig. 14.7. The figure illustrates the influence of oxygen at various partial pressures on the stress in the film after deposition and after annealing. Contamination severely changes the

Table 14.4 Standard heat of reaction $\Delta H°$ for the oxidation of a silicide to form a metal oxide and SiO$_2$ and its correlation to chemical stability. Strydom and Lombaard (1985). With kind permission of Elsevier

Silicide	Metal oxide	$\Delta H°$ (kJ (g atom O))$^{-1}$	Chemical stability		
			Theoretical[a]	Experimental (high temperature)	Experimental (low temperature)
ZrSi$_2$	ZrO	− 459	No	No	No
	ZrO$_2$	− 440			
TiSi$_2$	TiO	− 430	Yes	Yes	No
	Ti$_2$O$_3$	− 429			
	TiO$_2$	− 417			
CrSi$_2$	Cr$_2$O$_3$	− 393	Yes	Yes	No
	CrO$_2$	− 363			
	CrO$_3$	− 310			
NiSi$_2$	NiO	− 375	Yes	Yes	Yes
	Ni$_2$O$_3$	− 341			
CoSi$_2$	CoO	− 371	Yes	Yes	Yes
	Co$_3$O$_4$	− 358			
PtSi	PtO	− 288	Yes	Yes	Yes
	Pt$_3$O$_4$	− 254			
	PtO$_2$	− 232			

[a]If the most negative heat of reaction for a silicide is more positive -431 kJ (g atom O)$^{-1}$ (the heat of reaction for forming SiO$_2$ only) it can be theoretically expected to be chemically stable

properties of silicides. An effect of oxygen contamination on the properties of TaSi$_2$ films is shown below as an example.

NiSi$_2$ is considered a replacement of other silicides due to its lower processing temperature.

14.4 MoSi$_2$-Oxidation

There is a great need for high-strength, oxidation-resistant materials for elevated structural applications at elevated temperatures, particularly in aircraft gas turbines and in spacecraft. MoSi$_2$ has excellent oxidation resistance and some of its elevated temperature mechanical properties such as strength and creep resistance, thermal shock resistance etc. are of interest also in the re-entry cones in 'space-age' applications.

Thermal oxidation of silicides in wet and dry oxygen at 1000 °C is illustrated in Fig. 14.8. Oxidation of MoSi$_2$ is included. The excellent oxidation behavior even of

Table 14.5 Comparison of silicide oxidation kinetics. Strydom and Lombaard (1985). With kind permission of Elsevier

Silicide	Substrate	R_0 ($m^2 s^{-1}$)	Q (eV)	Oxide thickness[a] (Å)	Reference
Oxidation in steam or wet oxygen					
TiSi$_2$	Si (polycrystalline)	6.0×10^{-11}	1.51	4800	7
	Si (?)	2.0×10^{-11}	1.39	4800	7
	Si⟨100⟩	1.3×10^{-11}	1.4	3684	This work
PtSi	Si⟨100⟩	2.0×10^{-11}	1.4	4570	This work
CoSi$_2$	Si⟨100⟩	1.7×10^{-11}	1.4	4200	This work
	Si⟨111⟩	1.14×10^{-12}	1.1	4276	27
Crsi$_2$	Si⟨100⟩	1.6×10^{-11}	1.4	4087	This work
NiSi$_2$	Si⟨111⟩	4.3×10^{-13}	1.0	4150	11
	Si⟨100⟩	2.1×10^{-13}	0.99	3029	This work
TaSi$_2$	Si (polycrystalline)	4.8×10^{-12}	1.4	2250	10
WSi$_2$	Si (polycrystalline)	4.0×10^{-12}	1.0	12,600	6
	SiO$_2$	1.7×10^{-14}	0.35	15,800	6
Oxidation in dry oxygen					
TiSi$_2$	Si⟨100⟩	2.0×10^{-13}	1.3	720	This work
PtSi	Si⟨100⟩	2.3×10^{-13}	1.3	773	This work
CoSi$_2$	Si⟨100⟩	2.7×10^{-13}	1.3	827	This work
	Si⟨100⟩	9.8×10^{-13}	1.39	1056	27
CrSi$_2$	Si⟨100⟩	1.0×10^{-13}	1.3	510	This work
NiSi$_2$	Si (polycrystalline)	3.0×10^{-12}	1.5	1126	11
MoSi$_2$	SiO$_2$	4.1×10^{-12}	1.6	827	8

[a] After 1 h at 1000 °C

Fig. 14.4 Rutherford backscattering spectra of NiSi$_2$ silicide before (-) and after (⋯) thermal oxidation in steam at 900 °C. Strydom and Lombaard (1985). With kind permission of Elsevier

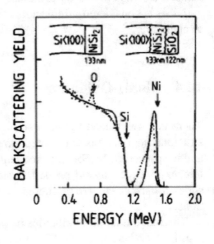

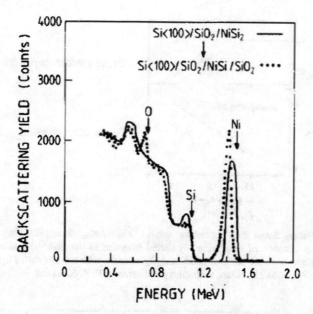

Fig. 14.5 Spectra showing steam oxidation of NiSi₂ on inert SiO₂ substrate. The spectra for NiSi₂, was measured at a target angle of 60°. Strydom and Lombaard (1985). With kind permission of Elsevier

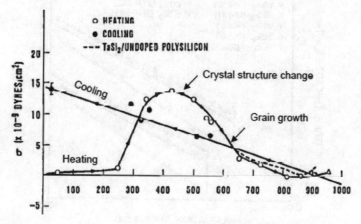

Fig. 14.6 The stress variation as a function of temperature. Physical Stress in Silicides. K. C. Saraswat 311 Notes, With kind permission of Prof. Saraswat

monolithic MoSi₂ is manifested in coatings sustaining very large number of cycles (454 cycles) in cyclic oxidation (Yoon et al. 2004). The high temperature oxidation resistance is due to the formation of a protective silica (SiO₂) film. However, the major problem with the use of unreinforced MoSi₂ for structural applications has been inadequate high temperature strength on applying the required stress above 1200 °C and the relatively low fracture toughness below the brittle to ductile temperature at about 1000 °C. Above this temperature MoSi₂ yields plastically by dislocation

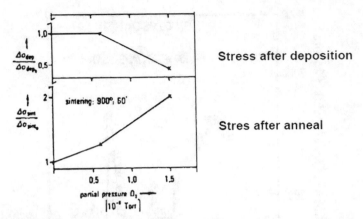

Fig. 14.7 The stress $\Delta\sigma_{dep}$ after deposition, stress change $\Delta\sigma_{sint}$ during sintering of evaporated TaSi$_2$ films as a function of increasing O$_2$ partial pressure in the residual gas. All values are normalized to those obtained without additional O$_2$, marked with a suffix "0". Physical Stress in Silicides. K. C. Saraswat 311 Notes, with kind permission of Prof. Saraswat

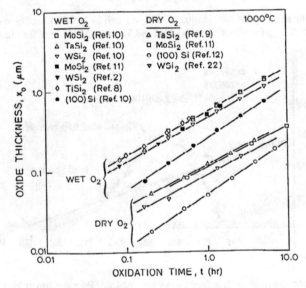

Fig. 14.8 Oxide thickness versus oxidation time for silicides. Interconnections: silicides. K. C. Saraswat 311 Notes, with kind permission of Prof. Saraswat

motion. Thus the need for strengthening by alloying or the use of composites is essential. Alloyed MoSi$_2$ and composites are out of the scope of this book. Despite the excellent oxidation resistance in air up to high temperatures (resulting from the formation a protective silica layer) during the consolidation technique of the powder, the formation of a glassy amorphous siliceous phase at grain boundaries degrade the mechanical properties at both ambient and elevated temperatures. Elimination of the glassy phase-which forms because of the affinity of the large surface area of the powder to oxygen-is achieved by addition of a deoxidant such as carbon. Carbon additions to oxygen-containing $MoSi_2$ powders improve the high temperature properties of the $MoSi_2$ by removal of SiO_2 (also by the formation of SiC). In Fig. 14.9 the hardness as a function of temperature is shown for $MoSi_2$ with and without C additions. The indentation resulting from the hardness test in $MoSi_2$ is seen in Fig. 14.10. Intergranular cracking is observed due to the hardness indentation. The fracture toughness performed in air is shown in Fig. 14.11. The fracture toughness of $MoSi_2$ is improved when the oxygen is removed from the $MoSi_2$ as seen in Fig. 14.11 by C additions. The toughness line representing $MoSi_2$ is below those related to the C added silicides. Thus C additions to oxygen containing $MoSi_2$ powders improve the high temperature properties by removal of the siliceous phase at grain boundaries.

Also the creep properties of $MoSi_2$ was investigated and the results are presented below. The consolidated alloy is referred to as the base alloy. It was fabricated by consolidation of commercial $MoSi_2$ powder. Hot pressing was used for the consol-

Fig. 14.9 The effect of carbon on the variation of hardness with temperature. Maloy et al. (1992). With kind permission of Elsevier

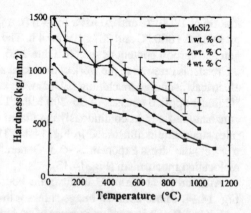

Fig. 14.10 Vickers microhardness indentation produced in $MoSi_2$, at 1000 °C. Maloy et al. (1992). With kind permission of Elsevier

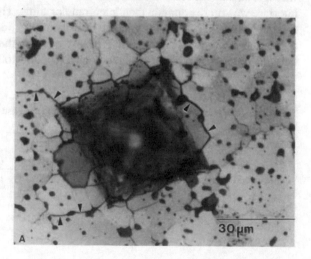

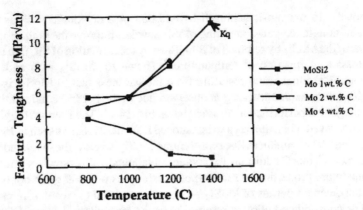

Fig. 14.11 The effect of carbon on the variation of fracture toughness with temperature. Maloy et al. (1992). With kind permission of Elsevier

idation at 1800 °C and 20 MPa for 4 h in a graphite die, followed by hot isostatic pressing at 1800 °C and 207 MPa for 4 h. The base alloy had a higher oxygen, nitrogen and carbon content as seen in Table 14.6. For the detailed experimental method the reader is referred to the work of Suzuki et al. (1993). The $MoSi_2$ specimens cut by electrical discharge machining (EDM) were compressive creep tested at 1050–1300 °C under applied stresses of 35–300 MPa. The experiments were conducted under constant load and in continuously circulated air. The steady state (minimum creep rate) strain rate is illustrated in Fig. 14.12. The data indicate a power law, namely, $\sigma \approx \dot{\varepsilon}^n$; the stress exponent is ~3.5. Generally n = 3 means a creep controlled by dislocation mechanism (Fig. 14.13).

After creep at 1200 °C in air grain boundary glass film is present as seen in Fig. 14.14. Phase-contrast image shows a thin amorphous film (less than 0.5 nm), which is difficult to resolve because of an inclination of the boundary. Glass film at interface was more common than intergranular films. The creep deformation mechanism of the $MoSi_2$ base alloy involves dislocation glide with climb being the rate controlling process. Accumulation of glassy phase at these sites during creep facilitates the process. Cavitation was rarely observed in $MoSi_2$.

Table 14.6 Chemical composition of the composite and the base alloy. Suzuki et al. (1993). With kind permission of Elsevier

Material	[Mo]/[Si] ratio		Interstitial elements			Reinforcement (vol.%)
	Mo (at.%)	Si (at.%)	O (wt.%)	N (wt.%)	C (wt.%)	
Composite	33.4	66.6	0.13	0.018	n.a.	29.1
Base alloy	32.2	67.8	0.65	0.100	0.376	–

Fig. 14.12 Steady state (minimum) strain rates versus applied stress for the base alloy in circulated air (Composite data are shown for comparison). Suzuki et al. (1993). With kind permission of Elsevier

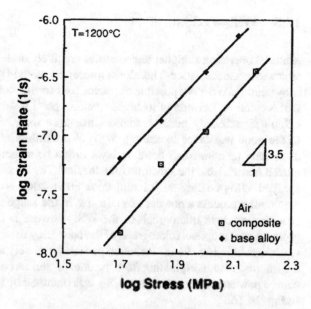

Fig. 14.13 Metallography of the MoSi₂ after etching with a modified Murakami's reagent (15 g $K_3Fe(CN_6)$, 2 g NaOH, 100 ml water). Suzuki et al. (1993). With kind permission of Elsevier

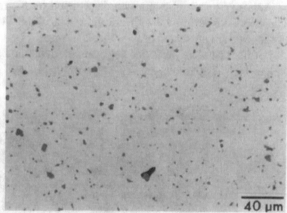

Fig. 14.14 Grain boundary structure in base alloy after creep in air at 1200 °C. Grain boundary in base alloy showing trace of glass film. Suzuki et al. (1993). With kind permission of Elsevier

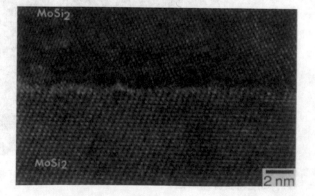

14.5 WSi$_2$-Oxidation

Material operating at higher temperatures are likely undergo oxidation and/or corrosion with scale formation. Therefore an accepted method to overcome these problems is by coating. WSi$_2$ is a possible candidate for forming corrosion and oxidation resistant protection. Because of its higher melting point it was hoped that WSi$_2$ coatings exhibit significantly better oxidation resistance than MoSi$_2$ coatings. In addition to the good resistance to scaling, WSi$_2$ also exhibits good resistance to corrosion as already mentioned. A bonding layer might be sometimes required between the coated material and the substrate for adhesion. The adhesion bonding for the plasma sprayed WSi$_2$ coating an 0.1 mm thick 80Ni–20Cr bonding layer was used. For the coating process a powder of grain size in the range 22.5–45 μm was used. The scanning electron micrograph of the WSi$_2$ powder is illustrated in Fig. 14.15. A typical lamelar optical micrograph of the base alloy used for the creep investigations is illustrated in Fig. 14.13. Inclusion particles probably of SiC and SiO$_2$ are seen in the microstructure originating from the unexpected carbon and oxygen present in the source powders. Structure of the WSi$_2$ was obtained by the plasma spraying as seen in Fig. 14.16.

Adhesive tensile strength was determined on thermally sprayed WSi$_2$ coatings. The tests were performed with and without the 80Ni–20Cr bonding layer. The results of the tests are presented in Fig. 14.17.

Further test were performed to evaluate dynamic behavior by means of rotating fatigue tests. The results are illustrated in Fig. 14.18.

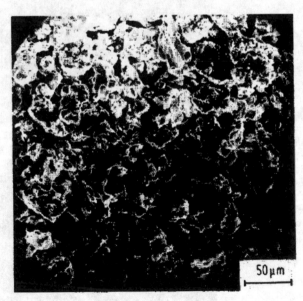

Fig. 14.15 Scanning electron micrograph of the WSi$_2$ powder used. Knotek et al. (1987). With kind permission of Elsevier

Fig. 14.16 Structure of WSi$_2$ coating. Knotek et al. (1987). With kind permission of Elsevier

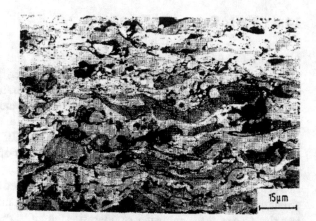

Fig. 14.17 Adhesive tensile strengths of WSi$_2$ coatings in the presence and in the absence of an 80Ni–20Cr intermediate layer. Knotek et al. (1987). With kind permission of Elsevier

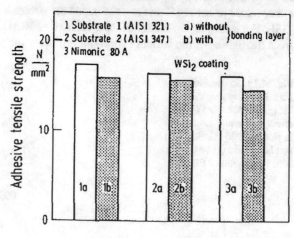

Fig. 14.18 S-N diagram of uncoated and WSi$_2$-coated Nimonic 80A. Knotek et al. (1987). With kind permission of Elsevier

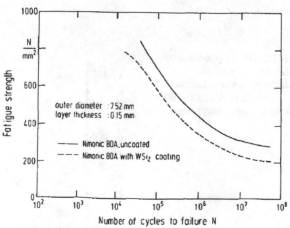

14.6   TiSi$_2$-Oxidation

Rutherford backscattering spectra of TiSi$_2$ before and after thermal oxidation in steam are illustrated in Fig. 14.19. As indicated earlier, Tables 14.1 and 14.2 show the chemical and structural stability at various temperatures, while in Tables 14.3, 14.4 and 14.5 the kinetics of oxidation, the enthalpies and activation energies for the reactions to form a specific silicide are indicated. Note, unlike in the case of CoSi$_2$ and NiSi$_2$ the Rutherford backscattering spectra of TiSi$_2$ indicates in addition to SiO$_2$ the formation of a metal oxide also, namely, that of TiO$_x$ during the oxidation reaction.

As in the majority cases alloyed silicides are presented in the literature because the strengthening effect of the various additives; no mechanical properties of TiSi$_2$ per se exist. One of the composites of interest is the SiC strengthened TiSi$_2$. From plots extrapolated to zero SiC one can estimated the strength properties of pure unalloyed TiSi$_2$. For example, from Fig. 14.20 a value of ~170 MPa at zero SiC can be read off from the figure for the bending strength. Similarly the yield strength can be derived

Fig. 14.19   Rutherford backscattering spectra of TiSi$_2$ silicide before (-) and after (···) thermal oxidation in steam at 900 °C. Strydom and Lombaard (1985). With kind permission of Elsevier

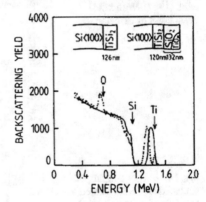

Fig. 14.20   Dependence of bending strength of SiC/TiSi$_2$ composites on SiC content. Li et al. (2000). With kind permission of Elsevier

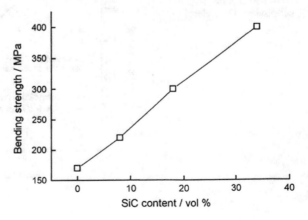

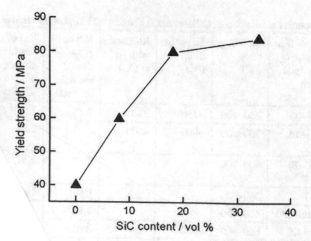

f yield strength of SiC/TiSi$_2$ composites on SiC content at 1200 °C. Li
nission of Elsevier

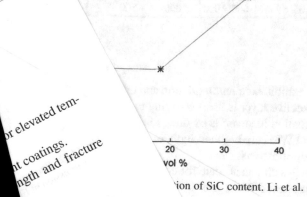

66
.67
0.64
0.59

or elevated tem-

t coatings.
ngth and fracture

ion of SiC content. Li et al. (2000). With

value of 40 MPa at zero SiC in
tained from the plot of 14.22
riments were conducted in

Eng. A **155**, 159 (1992)

electronics, where their
ncluding in this book

993)
35 (2004)

Table 14.7 Properties of silicides. K. C. Saraswat 311 Notes, with kind permission of Prof. Saraswat

Silicide	Thin film resistivity ($\mu\Omega$ cm)	Sintering temp (°C)	Stable on Si up to (°C)	Reaction with Al at (°C)	Nm of Si consumed par nm of metal	Nm of resulting silicide per nm of metal	Barrier height to n-Si (eV)
PtSi	28–35	250–400	~750	250	1.12	1.97	0.84
TiSi$_2$ (C54)	13–16	700–900	~900	450	2.27	2.51	0.58
TiSi$_2$ (C49)	60–70	500–700			2.27	2.51	
CO$_2$Si	~70	300–500			0.91	1.47	
CoSi	100–150	400–600			1.82	2.02	
CoSi$_2$	14–20	600–800	~950	400	3.64	3.52	0.65
NiSi	14–20	400–600	~650		1.83	2.34	
NiSi$_2$	40–50	600–800			3.65	3.63	0.
WSi$_2$	30–70	1000	~1000	500	2.53	2.58	0
MoSi$_2$	40–100	800–1000	~1000	500	2.56	2.59	
TaSi$_2$	35–55	800–1000	~1000	500	2.21	2.41	

Summary

- Silicides exhibit excellent oxidation and corrosion resistance
- SiO$_2$ protective layer is formed during oxidation of silicides
- SiO$_2$ formation in steam is by orders faster than in dry air
- MoSi$_2$ and WSi$_2$ are high-strength, oxidation-resistant materials f
  perature applications
- WSi$_2$ is a possible candidate for corrosion and oxidation resista
- SiC strengthened TiSi$_2$ improves bending strength, yield str
  toughness.

References

O. Knotek, R. Elsing, H.-R. Heintz, Surf. Coat. Technol. **30**, 107 (198
J. Li, D. Jiang, S. Tan, J. Eur. Ceram. Soc. **20**, 227 (2000)
S.A. Maloy, J.J. Lewandowski, A.H. Heuer, J.J. Petrovic, Mater. Sci.
.C. Saraswat, Physical Stress in Silicides. EE 311. Saraswat Notes
. Strydom, J.C. Lombaard, Thin Solid Films **131**, 215 (1985)
uzuki, S.R. Nutt, R.M. Aikin Jr., Mater. Sci. Eng. A **162**, 73 (
oon, K.-H. Lee, G.-H. Kim, J.-H. Han, Mater. Trans. **45**, 2

Index

© Springer Nature Switzerland AG 2019
J. Pelleg, *Mechanical Properties of Silicon Based Compounds: Silicides*,
Engineering Materials, https://doi.org/10.1007/978-3-030-22598-8

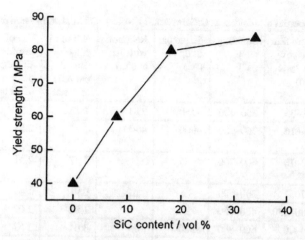

Fig. 14.21 Dependence of yield strength of SiC/TiSi$_2$ composites on SiC content at 1200 °C. Li et al. (2000). With kind permission of Elsevier

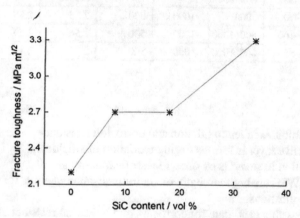

Fig. 14.22 Fracture toughness of composites as a function of SiC content. Li et al. (2000). With kind permission of Elsevier

from the plot of Fig. 14.21 to obtain an approximate value of 40 MPa at zero SiC in the material. The fracture toughness, K_{Ic} can also be obtained from the plot of 14.22 yielding a value of ~2.2 MPa m$^{1/2}$ at zero SiC. The experiments were conducted in air (Fig. 14.22).

Since many if not all of the silicides are used in microelectronics, where their resistivity is the property of prime interest, it was felt that including in this book some additional data summarized in Table 14.7 might be useful.

Table 14.7 Properties of silicides. K. C. Saraswat 311 Notes, with kind permission of Prof. Saraswat

Silicide	Thin film resistivity ($\mu\Omega$ cm)	Sintering temp (°C)	Stable on Si up to (°C)	Reaction with Al at (°C)	Nm of Si consumed par nm of metal	Nm of resulting silicide per nm of metal	Barrier height to n-Si (eV)
PtSi	28–35	250–400	~750	250	1.12	1.97	0.84
TiSi$_2$ (C54)	13–16	700–900	~900	450	2.27	2.51	0.58
TiSi$_2$ (C49)	60–70	500–700			2.27	2.51	
CO$_2$Si	~70	300–500			0.91	1.47	
CoSi	100–150	400–600			1.82	2.02	
CoSi$_2$	14–20	600–800	~950	400	3.64	3.52	0.65
Nisi	14–20	400–600	~650		1.83	2.34	
NiSi$_2$	40–50	600–800			3.65	3.63	0.66
WSi$_2$	30–70	1000	~1000	500	2.53	2.58	0.67
MoSi$_2$	40–100	800–1000	~1000	500	2.56	2.59	0.64
TaSi$_2$	35–55	800–1000	~1000	500	2.21	2.41	0.59

Summary

- Silicides exhibit excellent oxidation and corrosion resistance
- SiO$_2$ protective layer is formed during oxidation of silicides
- SiO$_2$ formation in steam is by orders faster than in dry air
- MoSi$_2$ and WSi$_2$ are high-strength, oxidation-resistant materials for elevated temperature applications
- WSi$_2$ is a possible candidate for corrosion and oxidation resistant coatings.
- SiC strengthened TiSi$_2$ improves bending strength, yield strength and fracture toughness.

References

O. Knotek, R. Elsing, H.-R. Heintz, Surf. Coat. Technol. **30**, 107 (1987)
J. Li, D. Jiang, S. Tan, J. Eur. Ceram. Soc. **20**, 227 (2000)
S.A. Maloy, J.J. Lewandowski, A.H. Heuer, J.J. Petrovic, Mater. Sci. Eng. A **155**, 159 (1992)
K.C. Saraswat, Physical Stress in Silicides. EE 311. Saraswat Notes
W.J. Strydom, J.C. Lombaard, Thin Solid Films **131**, 215 (1985)
M. Suzuki, S.R. Nutt, R.M. Aikin Jr., Mater. Sci. Eng. A **162**, 73 (1993)
J.-K. Yoon, K.-H. Lee, G.-H. Kim, J.-H. Han, Mater. Trans. **45**, 2435 (2004)

Index

A

Activation energy, 113, 114, 116, 118, 120, 121, 253
Activation volume, 46, 52, 53, 55, 89
Alloying, 212, 215, 218–220, 223–229, 232–241, 260, 264
Anomalous, 32, 44, 83, 220

B

Boron, 163, 165–166, 182, 175–179
Brittle, 17, 21, 28, 29, 34, 49, 123, 131, 133, 139, 142, 145, 147–149, 151, 153, 163–164, 170, 183, 194, 199, 211, 214, 216, 217, 225, 239, 241, 263
Brittle fracture, 49, 123, 148, 149, 217, 225
Burgers vector, 43, 53, 70–72, 74, 78, 80, 81, 83–87, 93, 97, 101, 104, 148, 193, 194, 196, 218, 220, 221, 224

C

Cleavage, 128, 131, 135, 145–147, 236
Climb, 69, 71, 88, 91, 97, 101, 107, 114, 117, 122, 222, 224, 253, 266
Composition, 1, 11, 15, 37, 39, 164, 166–168, 195, 212–214, 225, 227, 231, 234, 235, 237, 238, 255, 260, 266
Compression, 17, 21–22, 28, 31–34, 37, 43, 48, 64, 67, 70, 75–76, 83–84, 86, 97, 100, 110, 119, 129, 139, 141–144, 149, 196, 215–216, 219, 224–226, 235–236
CoSi$_2$, 1–5, 11, 17–19, 29–34, 67, 69, 73, 76, 78, 79, 81, 109, 119, 139–141, 163–164, 211, 255–260, 270
Cracks, 17, 26, 28, 60, 63–65, 123, 126–134, 145–148, 150–153, 164, 169, 183–190, 199, 201–203, 207, 210, 214, 226, 236–240, 255, 260, 265
Creep, 93, 107–121, 162, 166–167, 183, 191, 199, 210, 233, 243, 246–247, 251–253, 261, 265–267
Cross slip, 74, 143
Cubic, 5, 11, 29, 74
Cycle, 123, 124, 126, 130, 133, 136, 177–179, 263

D

Defects, 14, 19, 69, 71, 78, 224, 260
Deformation, 17, 21, 28, 32, 34, 43, 47, 49, 53, 58, 67, 71, 73, 83–85, 88–91, 93–94, 96, 97, 101, 107, 109, 114, 117, 124, 132, 141–142, 144, 148, 154, 155, 157, 166, 194, 215, 219, 220, 224, 253, 266
Diffusion, 3, 92, 107, 117, 122, 167, 169, 170, 177, 224, 232, 253
Dislocation, 69–102, 119, 147, 164, 193–199, 211, 219–224, 226, 229, 260
Ductile, 17, 21, 34, 123, 131, 134, 142, 164, 183, 199, 214, 263
Ductility, 32, 125, 139, 141–142, 154, 155, 163–166, 257

E

Endurance limit, 123, 138
Environment, 54, 126, 255

F

Fatigue, 123–137, 268
FeSi$_2$, 1–5, 8–11, 17, 25–28, 63–64, 155, 160, 211, 229, 231

© Springer Nature Switzerland AG 2019
J. Pelleg, *Mechanical Properties of Silicon Based Compounds: Silicides*,
Engineering Materials, https://doi.org/10.1007/978-3-030-22598-8